A Century of Flight at

Paton Field

A Century of Flight at
Paton Field

The Story of Kent State University's Airport and Flight Education

William D. Schloman and Barbara F. Schloman

The Kent State University Press Kent, Ohio

ISBN 978-1-60635-386-8
Manufactured in the United States of America

Cataloging information for this title is available at the Library of Congress.

23 22 21 20 19 5 4 3 2 1

The significance of the past lies with the future, while the present
is an indispensable connecting link between the two.

—Andrew W. Paton, "Aeronautics Training in Higher Education"

Contents

William Andrew Paton in front of KSU hangar (Photo by William D. Schloman)

Foreword

This impressive work by Bill and Barbara Schloman tells of the birth and life of Ohio's century-old Kent State University Airport and honors many people who have contributed to that life. My father was a principal among them. The formative years of KSU aviation coincided with my own.

Andrew W. Paton (pronounced "Payton") Field of the Kent State University Airport was named after my father, director of Aviation Technology Division at Kent State University during the division's early years from 1946 to shortly before his untimely death in 1964. Aviation and the KSU Airport were prominent topics around our dinner table. From an early age, my three brothers and I were thoroughly exposed to aeronautical subjects and experiences through our father.

His aviation classrooms and shop (called "laboratory" in this book) at Van Deusen Hall on Kent State's campus were integral parts of KSU aviation. They were located just four miles across town from the airport. As a young boy, I often walked to the shop after school to await a ride home at the end of my father's day. The cut-away inline, radial, opposed, and jet engines were fascinating for a mechanically minded child and served to educate me as well as his university undergrads. Skeletonized wings and fuselages were studied and rebuilt by his students, and my brothers and I frequently climbed into the bare cockpits to play pilot. The smell of fresh dope (aircraft lacquer) drying on cotton fabric always signaled the final dressing up of the reconstructed planes in preparation for their test flights. The Goodyear Duck, Republic Seabee, and especially the Waco open-cockpit biplane captured our interest the most. On special occasions, an all-too-brief "solo" in the shop's Link Trainer was a high point of a visit to my father's workplace.

But trips to the KSU Airport were always the best. We wandered around the huge Quonset-type hangar in awe, getting up close to so many airplanes. On those trips, there were occasional behind-closed-door meetings between Father and the airport contractor. We were never allowed to follow him behind those doors, and tension frequently filled the air after those sessions. On one unexpected, very early morning trip to the airfield in May 1948, Father showed us the Kent State University Flying Club's crumpled Aeronca

Chief still nosed into the ground in a patch of woods a short walk to the east of the hangar. It was a frighteningly impressive sight for a five-year-old boy, and I recall Father saying the flying club member shouldn't have been in the air. His passenger was badly injured and sued the club.

Sometimes Father took us for sightseeing flights around the area. Mother almost never came along, and we knew she worried excessively when any of us went flying, but we didn't! Special occasions induced Father to take one of us kids out of school for a day to fly to Columbus to shop for state surplus property for the university. Always a teacher, on those trips he imparted navigation skills and the rudiments of flying to us. In Columbus, we invariably landed at Don Scott Field, and Father was sure to tell us who Don Scott was and why Ohio State University's airfield was named after him.

The best flights with Father, however, were in the Waco. She was dolled up in KSU's Golden Flashes colors, with her shiny new yellow wings and deep-blue fuselage sporting its "golden flashes" of lightning streaking across its sides. To start the radial engine, a ground crewman threw his weight into the inertia starter's hand crank, slowly and gradually running its flywheel up to speed before Father would engage its clutch to turn the crankshaft over. The tandem open cockpits, staggered wings, and huge rumbling engine made flights in the Waco especially thrilling experiences for us as passengers and for our pilot too. A chandelle (steep climbing turn executed to gain height while changing direction) always capped those flights.

In late 1962, at the height of his career as director of KSU's Aerospace Technology Division and manager of the KSU Airport, Father had a grand mal seizure and was subsequently diagnosed with a glioblastoma, a very aggressive brain cancer. The fear he expressed then was not that this tumor might take his life (which it did, all too soon), but, as he said with tears in his eyes, "I'm afraid I may never get to fly again."

After his death in 1964 (and reminiscent of Ohio State University's Don Scott), the university attached Father's name to the KSU Airport. When I look up at the iconic, now historic, and still very active KSU hangar to see "Andrew Paton Field" written across its front in bright blue letters, I gratefully sense that Father *did* get to fly again.

The authors opine that Andrew Paton had a major influence on the development of the KSU Airport. This book conveys why and how he and many others contributed to making this important airfield what it is today.

In the early 2000s, when its very existence was threatened by university policy decisions, I and many other supporters joined Albert Beckwith's campaign to save Andrew Paton Field. The fact that the airport's salvation can be traced to the fortuitous decision by KSU president Glenn Olds in 1973 doesn't diminish our pride in having added our voices to the support effort.

With high interest, and much appreciation to the authors, I read this meticulously researched and interestingly informative chronicle of the airport to which Father dedicated so much of himself. May you also enjoy and learn from this volume produced by Bill and Barbara Schloman's most dedicated efforts.

William Andrew Paton, MD,
Orthopedic Surgeon, Indian Health Service, Ret.,
Anchorage, Alaska

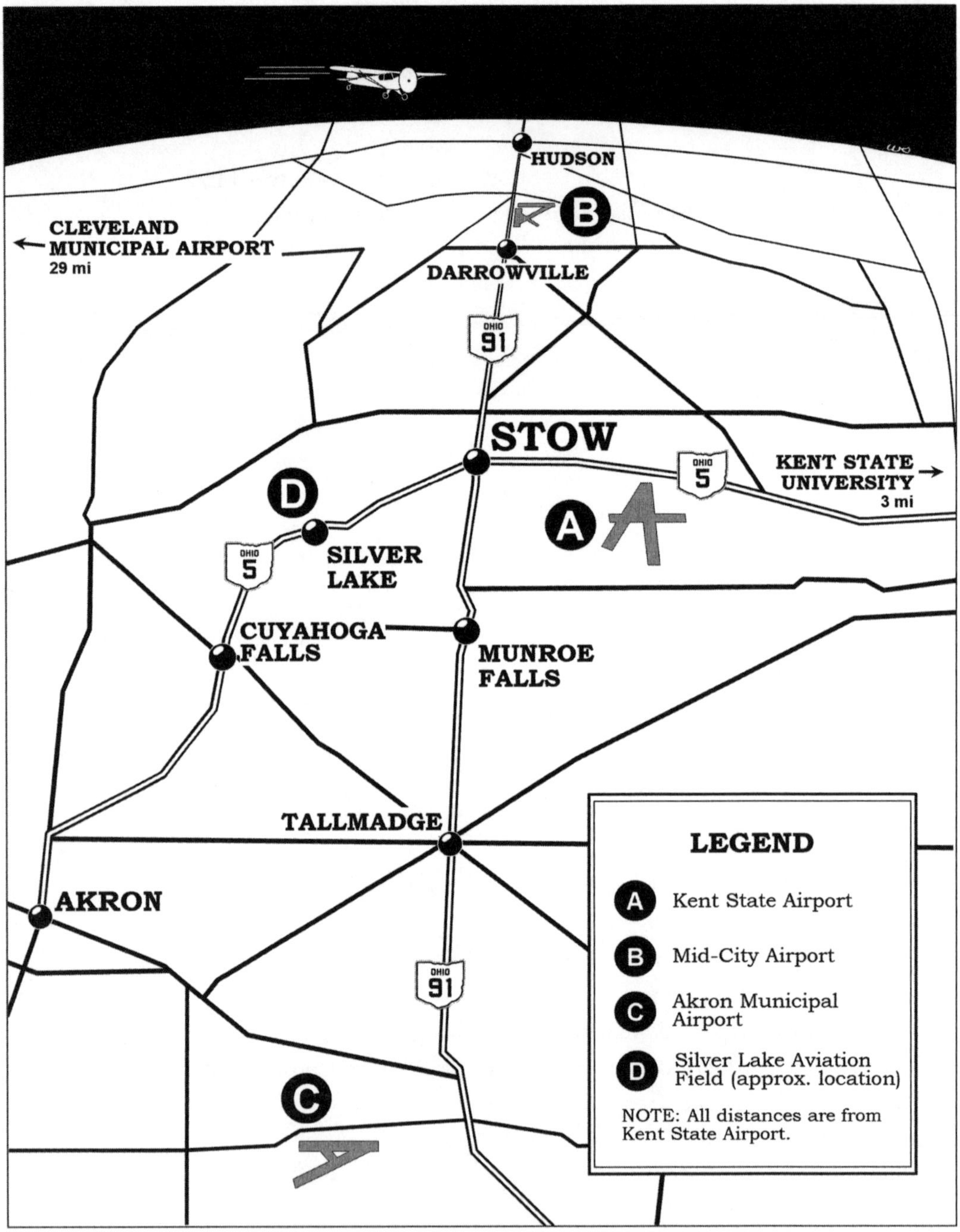

Area map showing the KSU, Mid-City, and Akron Municipal Airports (Map by William W. Schloman Jr.)

Preface and Acknowledgments

Our research on the history of the Kent State University Airport was undertaken to gather information for the airport's centennial celebration. As we uncovered stories of a rich and varied past, we also determined that the airport's longevity establishes it as the longest-surviving public-use airport in Ohio. Believing this history deserved a more complete telling led to this book.

What also became clear is that, for the first decades, this is a regional story that mirrors the development of aviation at that time. Activity at what was then Stow Field was intertwined in one way or another with Akron Municipal Airport (now Akron Fulton Airport) and Mid-City Airport in Darrowville (now Hudson). Therefore, while the early narrative centers on Stow Field, the other two airports are covered as well.

Many stories about the KSU Airport have been passed down over time. An important objective for us was to substantiate these or to provide clarification as needed. In this task, as well as for the entire project, we have benefitted from access to resources that were not available in the past. We have made every effort to document our sources to aid readers and future historians.

The many events making up the airport's story included some that were quite notable. But it was the people who set the path leading to the airport as we know it today that left a greater impression. We have been fortunate to connect with family members linked to the past and with individuals whose own history dovetailed with the airport's. They have been unfailingly generous in sharing stories, artifacts, and photos. We hope we have adequately represented their contributions.

From the Stow Field period, we wish to acknowledge Fran M. Hermance, Rick and Karolyn Gardner, Andrew Heins, Richard F. Smith, Carolyn Garrett Barton, Jean Schneider Kreyche, and Chris and Joe Van Devere. Our narrative highlights the Civil Pilot Training Program (CPTP), which led to Kent State University ownership of the airport. Our appreciation goes to Adrian Van Wyen for photos and artifacts of the period. We followed the lives of several of the first CPTP graduates and learned of their heroism in war and of the aviation careers of others. We were touched by their stories and by the willingness of their families to share. We wish to thank Laura

Assia, Russellyn "Rusty" Edwards, Donna Field, Ann Hendricks, Lowell "Buzz" Starner, Donald Stubbs, and Doug Wilkin.

Kent State University ownership of the airport has had its ups and downs. Fortunately, there were individuals who provided vision and leadership to set an early course. The family members of Andrew W. Paton, Marshall Garrett, Leonard Mack, Peder Otterson, and Richard Schwabe provided valuable information to help us understand that period. Our profound thanks go to William A. Paton Sr., Carolyn Garrett Barton, Michael R. Mack, Peder H. Otterson, and Jean Schwabe and family.

We extend special thanks to David Poluga, current airport manager, who started us on this project and provided support and information all along the way. We have been energized by the enthusiasm of alumni and their willingness to share photos and memories and assist with research. Our thanks go to Wilbur Biggin, Jonmichael Brogan, Sarah Deal, William Dedrick, Mike Everett, Mike Flanagan, Roger Hyatt, Stephanie Johnson, Scott Koeppl, Jack Kuss, Max Lovingood, Ted Orris, Clark Perrin, Ben Satyshur, Dave Schrank, William VonGunten, and Mark Wapennaar. Our appreciation extends also to others related to the airport's story, including Dennis Baden, Harriotte Beckwith Coke, David Dix, John Duncan, Lynn Feterle, Thomas Friend, Tom Grossman, Scott Layman, Scott Mills, Isaac Nettey, Tim Palcho, Jim Ripple, Charlie Wentz, and Eric Zimmerman. Our understanding of local history was aided greatly by the personal stories of Bob Adaska, Faith Cook, Stanley Crosier, Elvin Lichty, Fred Long, and Dale Platt.

We would be remiss not to acknowledge the role of Kent State University Libraries. The resources in Special Collections and Archives provided the foundation of our understanding of the Kent State University period. For their assistance in finding and helping us navigate those records, we are indebted to Amanda Faehnel and Cara Gilgenbach. Our thanks to Diane May and the staff in Interlibrary Loan. Other librarians also made a difference for us in this project and have our thanks. Beth Daugherty and Gretchen Quinn at Stow-Munroe Falls Public Library led us to unique local resources and provided useful historical background. Michael Sharaba at the International Women's Air and Space Museum went the extra mile for us in uncovering material. Laura Waayers, Naval History and Heritage Command, filled in some important details, as did Gwen Mayer, Hudson Library and Historical Society.

Friends were carried along on this journey, and we appreciate their encouragement. Finally, thanks to our families for living this story with us for the past three years. To Joanna, Billy, Jack, Bill, and John, your support kept us going. We will not forget this.

Chronology of Kent State University Airport Ownership and Operation

1889	Rev. Edward J. Smith begins buying property in Stow Township.
1917	Reverend Smith's property likely serves as a landing field.
1919–21	Ohio Flying School and Transport Company incorporates and operates airfield; goes into foreclosure in 1921.
1922–26	Reverend Smith and his son Fred regain control of Stow Field.
1923	Ralph F. Thomas leases Stow Field for the summer flying season.
1926	Safety Airways, Inc., headed by Gordon Lee of the Fageol Motors Company, leases Stow Field and lasts three months.
1927–32	Hugh C. Robbins establishes Robbins Flying Service at Stow Field. Robbins rebuilds hangar after 1929 fire and goes out of business in 1932.
1932–33	Allen T. Simmons incorporates Stow Flying Field with Frankie G. Renner as manager. The business ends in 1933 when fire destroys the hangar and planes.
1936	Stow Field is in foreclosure.
1937–39	Three investors lease Stow Field, building a new hangar and offering rides and instruction.
1940	Rudy Van Devere purchases Stow Field.
1942	Kent State University buys Stow Field; title is recorded in 1943. Rudy Van Devere retains ownership of the parcel with the hangar.
1945	Kent State University grants Rudy Van Devere exclusive commercial operating rights at the airport.
1947	Airport is classified as a general aviation field, given the identifier 1G3. The Kent State University Flying Club is founded, providing flight instruction.
1949	Ray Van Devere, Repair and Sales Depot, Inc., takes over the contract for field operations and management.
1953	Kent State University purchases property and hangars from Rudy Van Devere. Ray Van Devere continues under contract to manage the field and renames his business Van Devere Aviation in 1963.

1964 Kent State University Board of Trustees names the airport the Andrew W. Paton Airport; eventually named Kent State University Airport, Andrew Paton Field.

1966 Kent State University ends Ray Van Devere's contract to operate the airport and assumes full operational control, including the provision of flight training.

Chapter 1

The Early Years through 1922

Through periods of opportunity and luck that were more than matched with times of difficulty, Andrew W. Paton Field, home to Kent State University's airport, was part of the region's aviation narrative and managed to become the longest-surviving public use airport in Ohio.[1]

In its early years, the field was referred to variously as "Stow Field," "Stow Aviation Field," and "Stow Flying Field," and generically as an airfield, flying field, or landing field. The term "airport" did not come into common usage until the late 1920s and would not be applied to Stow Field's formal name until it came under Kent State ownership and was renamed Kent State University Airport in 1942.

In 1889, the Reverend Edward J. Smith, a local Methodist minister, began acquiring what was then farmland in Stow Township, Summit County. His first purchases were two parcels from James Season totaling 107.41 acres.[2] Over the next thirty years, Reverend Smith bought and sold acreage in these two parcels. His property was bounded on the north by Kent Road (Route 5, now Kent Road/Route 59) and on the south by today's North River Road.

Reverend Smith created what was considered a model farm that included water, gas, heating, and plumbing. Between 1908 and 1910, he built a large house and barn, making the property readily identifiable in the area. The *Akron Beacon Journal* society page called the home and grounds "palatial." Reverend Smith named his property "Kentview" because of its "magnificent view of Kent, especially the new buildings of the Kent normal school."[3]

Stow in 1910 was largely rural. The township had a population of 1,173, including the village of Silver Lake. The center of town was Stow Corners, located at the intersection of Kent Road and Hudson Road to the north (today's Darrow Road/Route 91). Interurban railway service connected Stow to Akron and Silver Lake Park to the west and to Kent and Ravenna to the east. The entire interurban rail system provided access to Cleveland to the north; Canton, Massillon, and New Philadelphia to the south; Wadsworth to the west; and Alliance and Warren to the east.[4]

Residents learned of the Wright brothers' historic flight at Kitty Hawk in 1903 from newspaper accounts, which were followed by stories of aviation

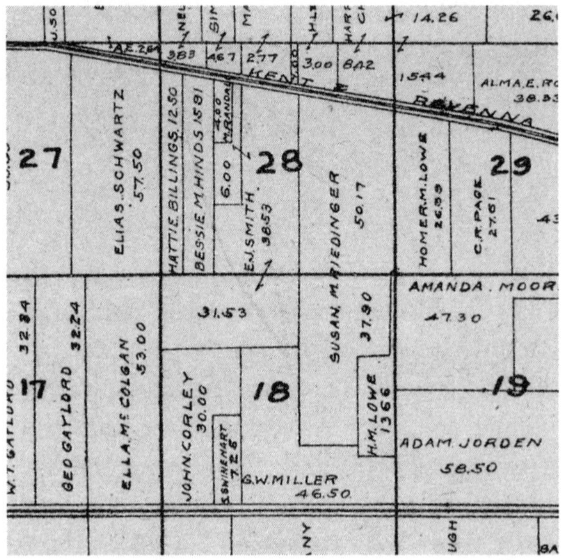

Right: Parcel map showing land owned by Rev. Edward J. Smith, 1910 (Atlas of Summit County, Ohio, 1910, Archival Services, University Libraries, The University of Akron)

Below: Rev. Edward J. Smith house on the property he called "Kentview," 1920 (Fran M. Hermance Collection)

Calbraith Perry Rodgers and his Wright EX airplane *Vin Fiz*. His landing in Franklin Township in September 1911 provided one of the first sightings of an airplane for residents of Portage County. Photo taken at the Chicago International Aviation Meet in August 1911. (Smithsonian National Air and Space Museum [NASM A-42906-A])

happenings from all over the country and abroad. But there was little opportunity for local residents to have any direct contact with these new machines. An Akron tailor in 1909 sought to provide some familiarity, however, by announcing through a newspaper advertisement that an exact working model of a Wright brothers' "aeroplane" could be seen in his display window.[5]

In fall 1911, residents anticipated the scheduled overhead flight of Calbraith Perry Rodgers, who was racing to be the first pilot to fly coast-to-coast in less than thirty days. Due to weather, he was forced to make an unscheduled stop on the Alonzo Johnson farm in Franklin Township west of Kent, drawing an excited crowd to see his flying machine. Rodgers became the first aviator to make a transcontinental flight and is recognized as an aviation pioneer for demonstrating the airplane's potential for travel and commerce.[6]

Early Airfields

This growing interest in flying prompted some businessmen to see the possibility of a commercial venture, leading to the establishment of the first airport in Summit County in 1912 at Silver Lake Amusement Park, which since 1874 had been a well-established recreational spot. In addition to a flying school, the plan was to manufacture "hydro-aeroplanes and aeroplane supplies."[7] For the July Fourth holiday in 1912, the park promoted its sixty-six-acre airport's aviation field and aviation exhibition, in addition to its two ballparks, thirty-six amusement departments, Chautauqua, and camping facilities.[8]

However, in 1913, the Silver Lake operation found itself challenged by three legal battles plus the death of a pilot in a biplane exhibition. In May 1914, another pilot died in an exhibition flight. By July of that year, the aviation business closed down, and all the equipment was sold. The hangars were converted into racehorse barns, and the airfield was turned into ballfields.[9] The loss of the Silver Lake Airport left an aviation void in the area. Locals would now be hard-pressed to experience anything aviation-related, while pilots needed an area landing field. Flying safely cross-country required suitable landing areas close enough together to permit stopping as needed, whether planned or due to mechanical or weather issues en route.[10]

Reverend Smith's property in Stow is believed to have served as an early landing field, perhaps as early as 1917. Although it has not been possible to substantiate this, it seems plausible given that farmers' fields were often used, and later descriptions called his field ideal for landing. A record in the Ohio Office of Aviation includes a note that the airfield predated World War I as a "military flying field," but it gives no details to authenticate this claim.[11]

Rev. Edward J. Smith (Richard F. Smith Collection)

Rev. Edward J. Smith (1842–1930), the original owner of the property that became the Kent State University Airport, was born in what is now West Virginia. He fought for the Union in the Civil War and suffered as a prisoner in several Southern prisons. Smith went on to receive several degrees, including a doctorate from Wesleyan University. For forty years, he served as a Methodist minister, with pastorates throughout northeast Ohio. When he moved to Stow from Chardon in 1900, he already owned land that he planned to use for general farming, with help from his two sons. Smith built an imposing thirteen-room house on the knoll on his property along with a large barn. He hosted many meetings at his home and was a familiar speaker at patriotic events. In 1906, he was widowed for the second time and, for the rest of his life, shared a residence with his son Fred. In 1919, with the lease of eighty acres to the Ohio Flying School and Transport Company, he relegated farming and airfield issues to Fred. Smith's first flight was on his eighty-first birthday, followed by many others with pilots from Robbins Flying Service. By 1930, he and Fred had moved to a smaller house on the property, where he watched airplanes taking off and landing from his room. At his death in 1930, Reverend Smith was known as "the flying parson."[12]

Aviation Post–World War I

Following World War I, aviation needed to demonstrate it was more than a tool of the military and could play a role in business and civilian life. The establishment of airmail service by the US Post Office proved to be that engine for development. The US Post Office inaugurated airmail service on May 15, 1918, between New York City, Philadelphia, and Washington, DC, with airplanes and pilots from the War Department. On August 12, the post office took over all airmail operations with its own pilots and established a well-run service with remarkable reliability. A transcontinental air route from New York to San Francisco was completed in September 1920. Lacking navigational aids and the needed instrumentation to fly after dusk, the mail was transferred to a train and then picked up at a later destination in the morning by plane. This air-rail combination was twenty-two hours faster than carrying solely by rail.[13]

The Air Mail Act of 1925, also known as the Kelly Act, was the first federal legislation affecting aviation. It authorized the US Postal Service to award airmail contracts to private carriers and set postal rates. The service was a boon to business, provided a safety net for the developing airline industry, and demonstrated the possibility of reliable air travel and transcontinental flight. Charles I. Stanton, an airmail pilot and later head of the Civil Aeronautics Administration, said of the airmail service: "We planted four seeds. . . . They were airways, communications, navigation aids, and multi-engined aircraft. . . . they are the cornerstones on which our present world-wide transport structure is built, and they came, one by one, out of our experience in daily, uninterrupted flying of the mail."[14]

At the end of World War I, thousands of military pilots who wished to continue flying returned to civilian life. Surplus military airplanes were plentiful and cheap, leading many young aviators to buy one and endeavor to make a living by moving from one locale to another, offering rides and performing stunts. When business became saturated in one place, they would move on to the next, earning them the initial label of "gypsy fliers" and later of "barnstormers." During the postwar period, Akron residents saw newspaper advertisements for "aeroplane exhibition flights" and "thrilling stunt flying."[15]

The Ohio Flying School at Stow Field

Interest in flying grew. That enthusiasm, plus an economy that had bounced back from a downturn following the end of the war, prompted a group of area men to begin a new aviation enterprise on Reverend Smith's property.

The Ohio secretary of state issued a charter to the Ohio Flying School and Transport Company in October 1919. Of the original investors, only vice president John F. Aston was a pilot, having served in the army air corps as chief instructor. The flying school's chief pilot was Lt. M. A. C. Johnson, also a flying veteran and former instructor at an army flying school in California.[16]

Reverend Smith's last land purchase in May 1919 increased the size of the future airport site to eighty acres. In spring 1920, the Ohio Flying School began erecting an eight-plane hangar in Stow Township, located on "the Smith farm property at Kentview, half way between Kent and Stow Corners." The field was to serve airplanes of all sizes making intercity or cross-country flights. The hangar, located on the east side of the field, was painted a bright orange to make it more visible to pilots. A service station at the field was to provide needed services for transient pilots. The *Kent Tribune* predicted, "Kentview field is destined to become a great flying field."[17]

The Northern Ohio Traction and Light Company agreed to make the field, located on the paved Kent-Ravenna Road, an interurban railway station stop, providing easy access to the new business. The airfield became Stop 53 (Aero-Station) on the Kent-Ravenna line.[18]

The Ohio Flying School purchased five "Jenny" airplanes (i.e., the Curtiss JN-4-D), a two-seater biplane used extensively in training pilots in World War I. The war surplus airplanes were ferried from Toronto to Stow Field by Aston and Johnson. Ex-army pilots with at least 400 hours of flying were recruited as instructors. The school gave successful students a proficiency certificate that enabled them to present their credentials to the Aero Club of America to take the required test to become licensed pilots.[19]

The enterprise also offered passenger and freight service, with plans to service Cleveland, Pittsburgh, Canton, and Youngstown. Passengers could arrange to be carried to "any city anywhere" at reasonable rates or take a pleasure ride over Kent, Akron, and other nearby towns. A 1920 newspaper ad promoted these "aeroplane flights" for only $10 (or $130 in 2019 dollars). In addition, the Ohio Flying School offered exhibition flying and aerial advertising and photography with the latter drawing interest from real estate companies.[20]

In its first year of operation in 1920, the Ohio Flying School introduced flying to the area and created visibility for its business by participating in a variety of events by doing aerial exhibitions at county and circuit fairs. In June, it participated in a celebration in east Akron held to recognize that community's business and civic developments. Attended by 150,000 people, the event included a two-mile-long parade that opened with aviation stunts performed by the Ohio Flying School. In July, the Associated Aviation Clubs

AEROPLANE FLIGHTS

July 5th, Only $10

To show the public the safety of sane flying, we have made a special rate for Monday, July 5th, of $10 per flight.

We have no "daredevil" pilots. Our flyers are former army instructors who have had over 1,000 hours in the air without accident.

Our planes are thoroughly inspected each day. We use the Curtis JN 4 planes in which 95 per cent of all United States army pilots were trained to fly. They are called the "Old Reliables" by pilots who have flown them.

We have four planes at our Stow Corners airdrome which already have made more than 1,000 successful flights.

Obey that impulse. Take advantage of th special low price and come early.

Flying Instruction

LOCATION OF FIELD

Stop 53, Kent-Ravenna car line. One mile beyond Stow Corners on Kent road.

Aerial Advertising

Ohio Flying School & Transport Co.

STOW CORNERS

Above: Field Day at the Ohio Flying School and Transport Company, Stow Field, summer 1920 (Dale W. Platt Collection)

Left: Newspaper ad for airplane rides at the Ohio Flying School and Transport Company, Stow Field (*Akron Beacon Journal,* July 3, 1920)

of Ohio sponsored Cleveland's first outdoor air show at Dunhamton Field, southeast of Garfield Park, which was attended by as many as 10,000 people. The Ohio Flying School participated and brought in Kent mayor Franklin Schmiedel to join other mayors in opening the show.[21]

At the very time that the Ohio Flying School was incorporated, John Gammeter, who earlier had been an important player at Silver Lake Airport, led area aviation enthusiasts in forming the Akron Flying Club. The club secured Wright Field, named for a local donor, and formed National Airway Service to provide offerings similar to the Ohio Flying School. Wright Field opened on July 6, 1920, with an aerial field day showcasing fifteen visiting airplanes, later considered as possibly the first air meet in Ohio.[22] With no field formally designated for the city of Akron, these outlying airfields would fill that role.

While activity for the start of the Ohio Flying School showed promise, the economy was hitting a rough spot. What has been called "the forgotten depression" began in 1920 and would continue into 1921. In November 1921, a restraining order was executed against the Ohio Flying School, and in December legal action was begun to foreclose on the school's mortgage. Similarly, the National Airway Service, Wright Field in Fairlawn, and the Akron Flying Club all disappeared by the end of 1920.[23]

In 1922, Reverend Smith and his son Fred purchased the Ohio Flying School and Transport Company buildings and the right to operate. Their intent was to support aviation by providing a landing field for visiting airplanes and cross-country fliers.[24]

Summary

The twenty-year period that began with the Wright brothers' flight in 1903 was a remarkable one that saw the nascent aviation era gain momentum with the development of aircraft and airways, a large and active group of daring pilots, and an increasingly engaged public. An entrepreneurial spirit brought the Ohio Flying School together with Reverend Smith and his landing field to form one of the first commercial airfields in Ohio. The enterprise gave Stow Field a short but credible start as an airfield during this early pioneering period.

Chapter 2

Akron's De Facto Airport, 1923–26

In 1920, Stow, Ohio, was an unincorporated township with a population of about 1,200, in contrast to that in the city of Akron with more than 208,000. A residential community, Stow had two churches and no industries. Half of the population lived within a one-mile radius of the school, which was located at the center of the town. Houses had access to electricity, telephone, and the sanitary sewer system but relied on wells for water. The area around Stow Field was rural, with numerous orchards.[1]

After acquiring control of Stow Field following the collapse of the Ohio Flying School and Transport Company, Reverend Smith's son Fred stated that he wanted to develop aviation in the area so that Akron would be recognized as a stop for airmail, freight, and passenger service. He would serve as operator for the next four years with the exception of a short stint in the summer of 1923, when the field was leased by Ralph F. Thomas, a former World War I aviator.[2]

When Fred F. Smith opened the flying season at Stow Field in April 1924, he announced that there were six airplanes available for rides and stunt exhibitions, including a new Waco biplane, and that instruction was available.[3]

Serving as Akron's Airfield

For much of the 1920s, Stow Field was the only established landing field in the area. As such, Fred Smith welcomed Akron-bound fliers. In April 1923, this included attendees at Akron's first aviation convention, including Brig.

Gen. William "Billy" Mitchell. Five Army Air Service Martin NBS-1 bombers arrived at Stow Field from McCook Field in Dayton, Ohio, each airplane with its own mechanic. Upon the group's arrival, Smith was asked to provide 150 gallons of gasoline for the airplanes' return trip.[4]

Famed World War I ace Maj. Eddie Rickenbacker also used Stow Field to make a speaking engagement in Akron in June 1923. Newspapers as far away as Indiana and California reported that Rickenbacker was able to start his morning with breakfast and his newspaper in Mt. Clemens, Michigan, then board his airplane *Spark Plug* at 9:45 A.M. and fly the 150 miles to Stow Field in one hour and fifty minutes. The details of such seemingly small events demonstrated the possibilities of aviation and fed the public's fascination with it.[5]

Stow Field received more formal recognition as the airfield for Akron with the federal government's mapping project in 1923. The project aimed to create reliable guides for landing fields and airmail routes. This was a significant undertaking as airmail pilots had little more than state road maps and notes made from personal observation to guide them. Lack of naviga-

Fred F. Smith (Richard F. Smith Collection)

Fred F. Smith (1874–1958), was born in Pennsylvania and raised in Chardon, Ohio. By 1910, he and his brother, Paul, were helping his widowed father on the Stow farm that would become one of the first established airfields in Summit County. Active in Republican politics, he was elected to his first term as a Summit County commissioner in 1920. A scandal and poor health led to his resignation at the end of 1925. Smith was proud that during his tenure, both the North Hill viaduct and the Summit County Home were built. He remained active in politics and was "credited by many with knowing more about Stow and Summit County history than any other resident." From the early 1920s until Stow Field was sold in 1940, Smith represented the family's interests. At times he was manager of the airfield, but he was always its promoter. By 1930, Smith and his family were living in a small house down the lane from the airport's big house. During the Depression, he worked as a laborer on a Works Progress Administration (WPA) road project and later at Goodyear Aircraft Corp., from which he retired in 1949. Smith was a longtime advocate for Stow incorporation, and he lived to see Stow become a village the year before his death. Smith died in the hospital ward at the Summit County Home on December 3, 1958, and was remembered as "one of the most colorful political figures in both Stow and Summit County history."[6]

Two army fliers with Fred F. Smith collecting landing field data for Stow Field in their De Havilland DH-4, August 1923 (Fran M. Hermance Collection)

tion guides, coupled with challenging terrain and weather conditions, often made flights treacherous. The mapping project had military and airmail pilots visit airfields to collect information for the creation of more descriptive guides that would include a graphical representation of the airfield, specific information regarding location and obstructions, and available services. Two army fliers visited Stow Field for that purpose in August 1923 in their large 12-cylinder DeHaviland DH-4 biplane bomber. They found the field to be satisfactory and asked that it be marked to increase its visibility from the air.[7] Markings were made with "STOW" on the barn roof and "AKRON" on the barn with a directional arrow.[8]

The *Aeronautical Bulletin* listed Stow Field under the "Akron" entry for Ohio and noted that electric car line or taxi service was available for access to Akron. A related series provided detailed route information by mile marker for select routes, including magnetic heading, the location of available landing fields, and landmarks. Stow Field was included in two of the published routes: Number 43, Columbus to Cleveland, Ohio; and Number 44, Cleveland, Ohio, to Moundsville, West Virginia.[9]

Safety Concerns

Aviation in the early 1920s was unregulated. Neither pilots nor airplanes were certified, allowing rogue "gypsy" fliers to perform irresponsibly and often in deteriorating military surplus equipment. The use of that wartime equipment also dampened the demand for new airplanes. In 1923, up to 1,200 surplus military airplanes were estimated to be in operation, often in the hands of operators with varying levels of competence. The industry's push for federal legislation was to no avail until passage of the Air Commerce Act of 1926. This law addressed pilot licensing, airworthiness of aircraft and major components, navigational facilities, and air traffic. It gave the federal government responsibility for fostering air commerce and making and enforcing safety rules.[10]

But in 1923, and lacking federal regulation, Akron along with several other municipalities passed an air ordinance regarding reckless flying. Akron's ordinance against stunt flying over the city was tested in November when a pilot and automobile dealer took off from Stow Field and circled Akron, dropping promotional leaflets for a car show. They were chased in the air by the city's air policeman, Lt. Harold A. Kullberg, who arrested them upon landing at Stow Field. This was believed to be the first arrest in the country for violation of air traffic rules. Beyond the notoriety he received for enforcing the law in the skies, Lieutenant Kullberg was a highly regarded World War I ace, having received the Distinguished Flying Cross for downing nineteen German aircraft while flying for the British Royal Flying Corps.[11]

To address the safety and business interests of commercial aviators, interested parties met in June 1924 and formed the Commercial Aircraft Association of Akron, thought to be the first organization of its kind in the country. The group formulated safety rules requiring inspection of all airplanes prior to taking passengers aloft. The association would also investigate crashes.[12]

The first fatal accident at Stow Field, and the most publicized, occurred in June 1924. Instructor Howard Calvert was flying with his student, Elwilda Wickersham. He had been an instructor at Stow Field for a year, and she had been taking lessons for three months. The student was at the controls when the Waco biplane stalled while entering the first turn of a "loop-the-loop." Unable to recover, the airplane fell a thousand feet to the ground, breaking the necks of both occupants. The Commercial Aircraft Association investigated and determined that the instructor's control stick had come loose from its socket and that the student had insufficient experience to right the airplane from an upside-down position. Fred Smith maintained that if there were no stunt flying, there would be no accidents.[13]

That summer, the Akron aviation community lost a leader when another accident occurred at Stow Field, claiming Lieutenant Kullberg, the recently named president of the Commercial Aircraft Association. A training flight with a student turned into a freak accident when Kullberg lost consciousness and slumped against the controls, causing the airplane to crash and killing both occupants. It was determined that Kullberg had suffered a case of sunstroke, which led to his loss of consciousness.[14]

Airplane Manufacture and Testing

In the 1920s, a substantial portion of the estimated two hundred aircraft manufacturers were individuals or small local companies, most of which would not survive the upcoming economic downturn. Stow Field was the site for airplane construction and testing for several of these entrepreneurs. One was Guthrie Clark and his Dispatcheur G-3, a single-seat biplane weighing only 660 pounds and capable of reaching a speed of 35 mph with its 40-horsepower motor. Lieutenant Kullberg tested it in 1923 and commented that the small airplane was "comparable to riding in a motorcycle after having driven an auto."[15]

Clark had been assisted by V. C. Babcock, a well-established aviator and inventor who came to Akron in the 1920s and became a familiar presence at Stow Field and in Akron aviation circles.[16] In 1923, Babcock worked at Stow Field with Advance Aircraft Company. There is no indication that this business was connected in any way to the one of the same name that would become the Waco Aircraft Company.[17]

In 1924, he formed Babcock Aircraft Company to build airplanes in Akron and then test them at Stow Field. Babcock's Teal airplane was a familiar site at the airfield in 1925 and 1926. The Teal was a one-person, open-cockpit biplane with a 90-horsepower Curtiss OX-5 engine. In April 1925, it set a new

Dispatcheur G-3, Stow Field, 1923 (Dale W. Platt Collection)

Right: V. C. Babcock's Teal at Stow Field. Photo by David A. Bashline, ca. 1925 (Courtesy of the San Diego Air & Space Museum)

Below: Babcock's LC-11 Cadet tested at Stow Field in 1928. The new monoplane featured a "taper wing" to reduce the possibility of stalling. Plans to produce the airplane at Stow Field did not materialize. (T. David Gibbons Collection)

Stow Field altitude record when the pilot took the Teal to 9,200 feet until forced down by the extreme cold and low oil pressure.[18] Babcock entered the Teal in the second Ford Reliability Tour, which was 2,600 miles long and had stops at thirteen cities. The trophy would go to the airplane that "carried the biggest load in proportion to its power, moved at a good rate of speed, and could 'get in and out of small fields fast.'" The Teal completed only the first three legs of the course before being grounded by engine problems.[19]

Event Hosting at Stow Field

The success of Stow Field as a business required ongoing promotion to engage the public's interest. Fred Smith's efforts to promote Stow Field were undoubtedly aided by the connections and visibility he enjoyed as a Summit County commissioner. He served a short term in 1921 and then was reelected for a four-year term. This position gave Smith the opportunity to influence local decision makers, as well as to connect with the public and promote aviation and his business.[20]

In 1924, in an effort to promote interest in aviation, youngsters were encouraged to pursue aircraft modeling, a popular activity at that time for boys. In May, schoolboys were hard at work on their models in preparation for the area's first competition, which was sponsored by the *Akron Beacon Journal* and the National Aeronautic Association. In support of this upcoming contest, Stow Field held two events. On May 23, an "aerial circus" of fifteen airplanes took off from the airfield to fly in a *V* formation over Akron and perform maneuvers. It was timed for the end of a school day so children could view it. The next day, students from Summit County were invited to Stow Field for an airplane exhibition. Believed to be the first school field trip in Ohio to view airplanes, 300 students examined airplanes up close and asked questions of pilots. Both boys and girls were invited, although the modeling contest was only for boys.[21]

In lieu of a municipal airport in Akron, Stow Field in June and July 1925 hosted two of the largest events held to date in Summit County. Both were two-day air exhibitions, thrilling crowds with aircraft displays and aerial stunts. The first big aerial exhibition on June 6 and 7 was sponsored by the Akron chapter of the National Aeronautic Association in support of "safe and sane" commercial aviation. This was reportedly the largest gathering in the history of Summit County, drawing more than 75,000 people for each of the two days of the meet. Highlights of the meet included Capt. Charles Nungesser, a French World War I ace credited with bringing down 105

Pilots at Stow Field preparing for an aerial circus over Akron, May 21, 1924 (Courtesy of the *Akron Beacon Journal*)

enemy airplanes; the arrival of the Zeppelin *Los Angeles,* which hovered for a half hour over Akron before heading for Cleveland; and the second annual model aircraft meet and competition, where 200 boys entered models they had built with guidance from their industrial arts teachers.[22]

The second big aviation event in the summer of 1925 was the Akron Centennial Air Meet on July 18 and 19. With no municipal airfield, the opening of Akron's weeklong centennial celebration was held at Stow Field. The first day featured lighter-than-air events—the Junior National Balloon Race and the christening of Goodyear's newest airship, the *Pilgrim.* The first of Goodyear's planned "air yachts," the *Pilgrim* was the world's smallest blimp. It was only a hundred feet long and could carry two passengers in addition to the pilot and mechanic. Most notably, it was the first commercial blimp to be filled with helium and would be the first one that Goodyear used for advertising.[23]

Following the ceremony, four large, 35,000-cubic-foot balloons took to the air along with the *Pilgrim.* The balloon race was the first for "bags" of this size. Participating were three local entries and one from the US Army. The winner, Goodyear IV, stayed in the air for 22 hours and traveled 223 miles.[24]

Heavier-than-air events highlighted the second day of the meet. An estimated 80,000 people gathered within a one-mile radius of the field to see the events. More than two dozen US Army Air Service airplanes performed formation flying. Parachute jumping included a triple jump from a Martin bomber at high altitude by the Army Air Service men from Wilbur Wright Field. The meet was under the auspices of the Akron chapter of National Aeronautic Association, which discouraged stunt flying, believing that it misled people to accept that all flying was unsafe. Even so, later in the day

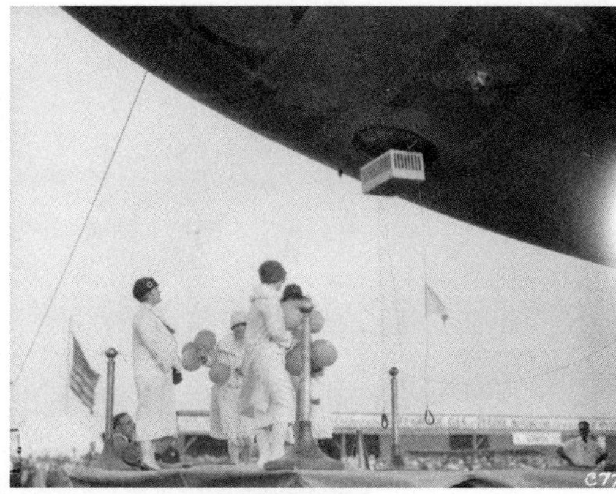

Above left: The *Pilgrim* over Stow Field, July 18, 1925 (Goodyear Photograph Series, Archival Services, University Libraries, The University of Akron).

Above right: Christening of the *Pilgrim* at Stow Field by Mrs. Paul W. Litchfield, July 18, 1925 (Goodyear Photograph Series, Archival Services, University Libraries, The University of Akron)

Left: The *Pilgrim* crew consisting of Capt. Jack Yolton, first mate Carl Wollam, and mechanic Albert C. O'Neil at Stow Field, July 18, 1925. Today, the gondola is in the National Air and Space Museum. (Goodyear Photograph Series, Archival Services, University Libraries, The University of Akron)

the airplanes from Wilbur Wright and McCook Fields in Dayton took to the air "with a skill that caused thousands to stop breathing as the airplanes went into nose dives, turns, and twists at low altitude." Commercial pilots provided rides throughout the day.[25]

Precautions were taken to protect attendees. B. E. "Shorty" Fulton, who would later play a key role in the creation and management of Akron's municipal airport, headed the safety crew that built rope fences and installed safety markers. The only accident at the meet was when Babcock's Teal biplane had rudder problems upon landing, causing its wings to knock down several bystanders. To avoid the traffic congestion that had plagued the June event, routes were better marked and two hundred special deputy sheriffs dispatched. Reportedly, all available parking spaces were filled for one mile around the field, and "the fields were dotted with the crowd that almost

Balloon race starting over Stow Field, July 18, 1925 (Goodyear Photograph Series, Archival Services, University Libraries, The University of Akron)

equaled half the population of Akron." It was estimated that ten thousand automobiles were accommodated for the lighter-than-air event on Saturday and twenty thousand for the heavier-than-air event the next day. Nonetheless, there would be critical letters to the editor about the rerouting of traffic and calling out Smith as commissioner and airport owner.[26]

Legal Troubles for Fred Smith

For Fred Smith, 1925 presented acute ups and downs. Stow Field business appeared healthy, with eight hundred flights covering five thousand miles, carrying one thousand passengers, and providing instruction and aerial photography.[27] In addition, the big events at the airport in June and July had been successful.

However, it was Smith's position as Summit County commissioner that caused particular strain. In January, a state investigation into a large county roads project charged him and others with negligence and laxity in safeguarding taxpayer interests. A special grand jury brought indictments in April

1925, charging bribery and misconduct. He and another commissioner were given the option to resign rather than face the charges. The other commissioner chose to quit and later, reportedly in poor health, leapt to his death from the North Hill viaduct. Smith declined to resign, challenged the legality of the special grand jury, and filed a motion to quash the indictments. The special grand jury, believed to have been handpicked by the prosecutor, was known as the "Million Dollar" grand jury, as it was comprised of prominent industrial and business leaders of Akron, including the Exalted Cyclops of the Summit County Ku Klux Klan.[28]

These legal threats came in the first months of the year, just as planning for the two big air meets was taking place. It may have been the pressure from this controversy that led Smith to put the airfield up for sale in April. A classified ad in the *Akron Beacon Journal* read: "Stow Aviation Field; 30 acres of as nice lying land as can be found any place, located right on the car line and paved road between Stow and Kent. This is a wonderful piece of property, suitable for good homes or for an aviation field, the purpose for which it is now used. The price is $15,000. This is cheap and should move readily."[29] The classified ad only ran twice, and no buyer came forward. The price of $15,000 would be equivalent to more than $218,000 in 2019.

On July 17, the day before the big centennial air show began, the judge ruled that the grand jury had been illegally constituted and convened. The indictments against Smith were thereby judged invalid. In October, the Summit County Board of Commissioners passed a resolution calling for Smith's resignation, citing misconduct in office due to intoxication at meetings. This led to his resignation as a county commissioner effective January 1, 1926. Smith's wife had urged this decision, noting that he had been under "a severe nervous and mental strain."[30]

Safety Airways

Given that Smith's health was stressed and that he had already contemplated separating himself from the airfield, it is not surprising that he sought an operator to take over. A group of California capitalists came forward in May 1926 to lease Stow Field for five years. The new firm was Safety Airways, Inc., and its president was Gordon Lee of the Fageol Motors Company, manufacturer of "Safety Coach" buses. Despite the similarity in names, there is no known connection between the two businesses.[31]

Safety Airways planned to create a successful commercial presence for heavier-than-air "machines." This would include the manufacture and sale of airplanes on the field, using the eighty acres of the Smith farm, plus an additional twenty-five acres from each of the two adjoining farms. The airplane to be built and sold at the field was the Lee Skipper, which had a cruising speed of 100 mph. The owners also planned to provide landing and hangar facilities for airplanes coming to the Akron area, as well as aerial photography and advertising services. The Cleveland-to-Louisville airmail route was due to begin soon, and the new field operators extended an invitation for those airplanes to stop for free at Stow Field.[32] The aviator in charge was Hugh C. Robbins, who would be a principal figure at the field for the next five years. Other aviators joined the effort, including Lt. Alfred Jung, a decorated German pilot of World War I.[33]

Newspaper ad for a Night Flying Circus at Stow Flying Field, July 30, 1926 (Courtesy of the *Akron Beacon Journal*)

In the flying season of 1926, Stow Field offered airplane rides and special aerial events on the weekends. Parachute jumping was popular fare and sometimes offered stunts that went beyond just jumping from height—such as jumping from the wing of a bomber or making a descent with two parachutes. Stow Field in June and July saw two former army parachutists endeavor to set the record for a delayed opening of the parachute. One was Sgt. Paul A. DeWeese. On June 28, 1926, he set an unofficial record for a delayed parachute jump, falling more than 1,100 feet before opening his parachute. DeWeese's airplane was piloted by Robbins to a height of about 2,200 feet. Nearly ten thousand witnessed the event. Later in the summer, Capt. P. G. Smith, the "Human Bullet," would attempt to break the world's free-fall record, which was then at 1,500 feet, but he managed only 1,400 feet before floating the remaining 1,800 feet to the ground.[34]

On August 8, 1926, Stow Field promoted a spectacular "night flying circus" with "airplane fireworks display supreme." Lieutenant Jung and local pilot Milo Shorts staged a night flying exhibition in front of a crowd estimated at four thousand to seven thousand people. With flares attached to the fuselage, they took off to execute the "flaming spiral," in which the airplane performed mock battle maneuvers and appeared to be in flames when making a sudden drop from a great height as if it were being shot down.[35]

Reportedly, fireworks were dropped from the airplane as bombs were being set off in the field below. The crowd did not immediately realize that Jung's airplane was in trouble when it came spiraling down and then crashed into the ground, causing him to be thrown free and fracturing his skull. He was critically injured and feared near death, but he did recover some months later and returned to Germany. Copilot Shorts suffered burns when fuel ignited upon impact. The cause of the accident was not determined.[36]

Following the accident, there was little mention of Safety Airways. The last was an account of an accident on August 28, when one of its pilots experienced an engine failure, causing him to fall a hundred feet and crash into an orchard fence not far from the field. A month after this incident, Safety Airways was gone, having lasted only three months. The next mention of Stow Field came in October, when local aviator J. W. Esch advertised special rates for rides that he would be giving on Saturdays and Sundays. Fred Smith's hope for a stable operator at Stow Field had not materialized.[37]

Interest for an Akron Municipal Airport

For much of the 1920s, the major aviation issue in the area was Akron's need for a municipal airport. The Manufacturers Aircraft Association maintained, "The spirit of civic progress, if not of common forethought, should urge every important community to have its municipal airdrome." The Air Commerce Act of 1926 made it clear that responsibility for establishing an airport was a local one. The act adopted the "dock concept," which maintained that airports were comparable to docks, which were built and maintained locally. The airways were analogous to the river and harbor improvements, for which the federal government assumed responsibility. Therefore, federal aid could not be used to build airports, but it would be available to maintain the airways.[38]

Yet across the country, as was the case in Akron, some local officials did not see a city airport as a municipal government responsibility.[39] Beyond civic pride or future commercial potential, there were two immediate, practical reasons for Akron to have an airport—landing an airmail stop for the city and convincing Goodyear Tire and Rubber Company that the area could support the needs of its growing airship enterprise.

As early as 1920, Scott Worley, president of the Ohio Flying School, and J. A. Beck, of the Akron Flying Club, had conferred with postal authorities in Cleveland about obtaining an airmail stop for the Akron area. Cleveland was a stop on the transcontinental airmail route from New York to San Francisco. Akron, in particular the rubber companies, embraced the service, finding that it reduced the time needed to send mail to San Francisco by four days.[40]

There was some hope in 1924 that postal inspectors might consider an Akron location for an airmail station. The initial station at Glenn L. Martin

Field in Cleveland was deemed inadequate due to the adverse weather conditions caused by proximity to Lake Erie, and a new site was being sought.[41]

Cleveland, however, won out with the selection of a thousand-acre site in Brook Park that would also serve as the city's municipal airport. Cleveland Municipal Airport opened on July 1, 1925, the day of the inaugural US Air Mail night flight into Cleveland. Following the opening of service at the new Cleveland Municipal Airport, the *Cleveland Plain Dealer* published the airmail schedule of airplanes landing at "Stow Field, Akron" and then returning to Cleveland.[42] There is no indication that this service continued beyond several months. Over the next years, Stow Field served as an emergency stop for airmail airplanes.

In addition to airmail service, the other major driver for an airport for Akron was tied to the needs of Goodyear. In late 1923, residents anticipated that the area would see tremendous growth from a developing airship industry. Goodyear had acquired patents from the German dirigible maker Zeppelin and expected to manufacture airships for the government and for the private market, employing up to five thousand workers. Goodyear judged its facilities at Wingfoot Lake as inadequate and undertook a survey of suitable fields close to Akron. Again, Stow Field was under consideration. Not surprisingly, it had support from Smith, who commented that both airplanes and dirigibles found that the field allowed for ease in making landings. The proposal gained the support of the Kent Chamber of Commerce.[43]

Although Goodyear expressed interest that land midway between Akron and Cleveland could serve for airmail, commercial airplanes, and the new Zeppelin works, a municipal airport for Akron was still two years away.[44]

Summary

The short period between 1923 and 1926 witnessed the needs of aviation mature from an emphasis on identifying available landing fields to establishing municipal airports. But a municipal airport serving Akron would face more struggles before becoming a reality. Stow Field received passing consideration, but its immediate problem as 1926 ended was to regain stability in its operation.

Chapter 3

Years of Promise, 1927–29

The Air Commerce Act of 1926 resulted in regulations requiring all aircraft engaged in interstate or foreign commerce to be licensed and the pilots of those aircraft to hold private or commercial licenses. It also laid out operational and air traffic safety rules. Airports were certified based on their suitability for use by registered aircraft. This included classification by size, equipment, and the extent to which night flying was supported. These regulations prompted Fred Smith early in 1927 to announce his intention to sell Stow Field because of the requirements necessary for an airfield to qualify as a first-class flying field.[1]

Smith was correct that Stow Field would not qualify for a high rating since it was lacking in several factors, including lighting for night flying. But it did meet all the minimum requirements to qualify for basic certification. In April, Smith announced he had leased the field to Hugh C. Robbins, who would operate under the name Robbins Flying Service.[2]

Robbins Flying Service

Hugh Robbins was an accomplished pilot, having done both barnstorming and commercial flying for over seven years. He had served as chief pilot for Safety Airways for a short period at its inception, offered passenger service at Fulton Field, and returned to Stow to establish his own business.[3]

Robbins Flying Service stated it would adhere to the new air regulations, including airplane inspections and the hiring of licensed mechanics and

Hugh C. Robbins (Courtesy of the International Women's Air & Space Museum, Cleveland, Ohio)

Hugh C. Robbins (1903–1987) was an Ohio boy with an early passion for flying. He tried to join the air corps at the onset of World War I, but he was rejected for being too young. In 1918, Robbins bought a "Jenny" biplane and took flying lessons, followed by an eighteen-month barnstorming tour. By 1921, he returned to Ohio before heading out again in 1923 with other pilots on his own aerial circus barnstorming tour. He was at Stow Field for a short time in 1925 and left to become a pilot for the Waco Aircraft Company in Troy, Ohio, beginning a long association with the company. Robbins returned to Stow Field to establish Robbins Flying Service in 1927. Existing through 1932, it was the longest-lasting business at the airfield prior to Kent State University ownership. After the Depression ended his business at Stow Field, Robbins pursued many aviation ventures. He was the Waco distributor at Cleveland airport, then worked with a firm to manufacture airplanes using "plywood plastics." He sought inventions for manufacture and merchandising and trained pilots to skywrite using charts with instrument flying. Always proud of being a safe pilot and operating within regulations, it was surprising that in 1941 he was disciplined by the Civil Aeronautics Board for flying too low over a British battleship as it entered New York Harbor. "[I] always look for trouble, but hope I never find it." At the time of his death in 1987 at eighty-five, Robbins's accomplishments included obtaining two US patents.[4]

pilots. It would provide instruction, transport service, aerial photography, and Akron's first airplane sales service. Robbins planned to make the field "one of the best flying bases in northern Ohio"[5] and stated there would be "no spectacular flying."[6]

In the updated *Airway Bulletin* for Stow Field published in 1927, F. F. Smith was listed as owner and Robbins Flying Service as operator. The airport map provided more detail from the previous edition, as did the description of the field's features and services, including a 30 x 150 ft. wooden hangar with a 16-foot clearance. Lacking were lights of any type, as well as available radio or telegraph communication.[7]

Robbins's business got off to a good start in the summer of 1927. Sundays offered the chance to engage the public by providing aerial entertainment. One stunt was a "double parachute drop" in which the aerial acrobat leapt from the airplane piloted by Robbins, opening his first chute. After descending several hundred feet, he cut the parachute loose and dropped through the air again before opening his second chute. There was also wing-walking

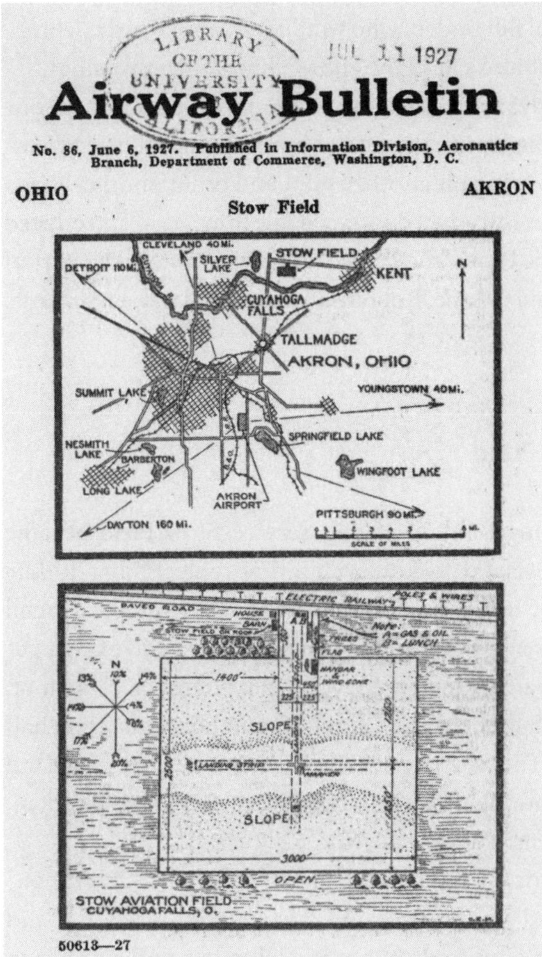

Left: "Ohio-Akron-Stow Field," *Airway Bulletin*, no. 86 (June 6, 1927). Field detail shows hangar on east side of the landing strip, hazards, and fuel available. (Courtesy of HathiTrust, https://hdl.handle.net/2027/uc1.b3122367?urlappend=%3Bseq=183)

Below: Robbins Flying Service hangar and Waco biplanes, Stow Field, ca. 1928 (Collection of Rick & Karolyn Gardner)

performed by local daredevil Bill Boden, who would thrill the crowd by climbing out of the cockpit of Robbins's airplane and walking along the wings.[8]

Airplane rides were always a big draw, and Robbins reported that for one Sunday in June he was in the air for almost ten hours giving rides. Increasing activity led Robbins to bring on another pilot and order another Waco biplane. The events and rides introduced more people to aviation and enticed some to want to learn to fly. Providing flight training became a key part of the operation, and, by the end of June, Robbins reported that sixteen students were taking instruction.[9]

Airline Service at Stow Field

Just one month after starting Robbins Flying Service, Stow Field became Akron's stop on the new Cleveland-to-Louisville airline route. Embry-Riddle Company was the operator of the new line and was vying for the airmail contract for this route. Passengers paid approximately twenty-five cents per mile. Freight and express parcels were charged by the pound or cubic foot. Using Waco biplanes, the flight between Cleveland and Louisville took half the time of the fastest train. The scheduled 2:00 P.M. departure from Cleveland continued on to Akron (Stow Field), then to Columbus-Dayton-Cincinnati, with a scheduled 11:40 P.M. arrival in Louisville, Kentucky.[10]

The five airplanes used for the inaugural trip on May 19, 1927, from Louisville to Akron (Stow) and on to Cleveland encountered heavy fog and drizzling rain, causing one pilot to drop out in Columbus and another to make a forced landing near Akron. The lead airplane experienced a delay when the pilot mistook Massillon for Akron in the heavy fog. Robbins did

Pilots and civilians greeting the area's first passenger airline service on its maiden trip on May 19, 1927. Rev. Edward J. Smith is eighth from the left, next to Hugh C. Robbins in the white flight suit. (Courtesy of the *Akron Beacon Journal*)

safely pilot his Waco to Stow Field and on to Cleveland. For the return trip that day from Cleveland to Louisville, one hundred Cleveland merchants were encouraged to send packages of their products to advertise the city and support the new line. But the airline service was short-lived, and by September it was discontinued after Embry-Riddle determined the route was unprofitable. Over the four months of operation, the line had carried 109 passengers and 380 pounds of express with only three forced landings.[11] Stow Field had provided the Akron area with its first airline service.

Charles A. Lindbergh

At the very time that the area's airline service through Stow Field was initiated, aviation experienced a momentous event with Charles A. Lindbergh's nonstop, solo airplane crossing of the Atlantic. Lindbergh's flight "gripped the consciousness of the entire civilized world."[12] Interest and confidence in aviation soared, and Lindbergh would be revered as the symbol of its promise.

Following his feat, Lindbergh traveled the country as an ambassador of aviation and supporter of municipal airport construction. One of his first visits was to Cleveland on August 1, 1927, where forty to fifty airplanes gave him an air reception as he approached the city. Two airplanes from Robbins Flying Service were part of the contingent, including Robbins in his Waco biplane. The next day, Robbins and his Waco met Lindbergh in the air again as the famed aviator circled over Akron. As Lindbergh passed over the city, he dropped a message with his greetings that fell to the street in front of the courthouse.[13]

Robbins as Aviation Promoter

Robbins's frequent presence at aerial events throughout the area was well covered by the newspapers, giving his business free promotion. Appearances were many and varied, including flying for Cuyahoga Falls' cleanup drive and the city of Orrville's dedication of the Lincoln Highway. In December 1929, his plane was illuminated, becoming the "Flying Christmas Tree" to promote outdoor holiday lighting.[14]

Robbins's business also distinguished itself by being the first in the area to sell airplanes through his Waco dealership. He had a long association with the company, going back to working for them as a pilot in 1924. He became the area dealer for the popular airplane in 1926 and by 1928 was the distributor for eighteen counties in northern Ohio. In 1929, he exhibited for Waco at the first of the Cleveland National Air Races.[15]

Akron's Struggle for a Municipal Airport

In 1927, the *Cleveland Plain Dealer* reported that "municipal airport fever is sweeping the country" and attributed it to enthusiasm following Lindbergh's flight. Akron surveyed other cities and found that over three hundred now had a municipal airport, a sobering finding given that Akron was the country's twenty-seventh largest city.[16] Akron leaders were waiting for the US Navy to give final approval to Goodyear-Zeppelin to build a "super-Zeppelin" for which Akron would be the likely assembly site. This in turn would influence the location of a future airport.

Akron's mayor formed a committee to identify a site suitable for both lighter-than-air and heavier-than-air craft and to explore possible financing. Initially, seven tracts of land were offered to the committee. The largest tract was in Darrowville, "a one-street town from Barlow Road in Hudson to beyond Fishcreek Road in Stow." Goodyear had optioned the property two years earlier. B. E. "Shorty" Fulton put forward Fulton Field on Massillon Road, southeast of Akron. Stow Field was no longer under consideration.[17]

The committee's first report in August 1927 presented a reduced list of four sites with a recommendation for Darrowville. It was characterized as the most adaptable site with a 1,100-acre level tract, ten miles from Akron. Fulton Field had 500 acres and with a location 3.7 miles from Akron was deemed the most accessible. To move forward, Akron City Council placed a municipal airport bond issue for $900,000 on the November ballot but without a final site decision.[18]

The campaign for passage of the bond issue included Continental Air Lines, the airmail contractor for the Cleveland-to-Louisville route. The company maintained that direct airmail service to Akron was contingent on having an adequate airfield. Distinguished aviator Clarence D. Chamberlin, holder of transatlantic records made only two weeks after Lindbergh's first flight, flew from New York to Stow Field to speak in support of the bond issue at the Akron Armory, followed by a radio broadcast.[19]

Despite these efforts, the bond issue was defeated in the November 1927 election. The next month, the navy gave Goodyear a contract for construction of two dirigibles for eight million dollars. Goodyear indicated its preference for Akron as the construction site, the only objection being its lack of airport.[20]

With the 1927 defeat of Akron's bond issue for a municipal airport, Robbins saw an opportunity to fill the void. The day after the election, Robbins, along with Allen T. Simmons and Donald Gottwald, took out incorporation papers for Akron Airport, Inc. The new venture was financed with an investment of $10,000. The plan was to expand operations at Stow Field with the purchase of new airplanes and to establish regular passenger routes from Akron. Robbins announced that he would construct additional hangars to deal with the expected increase in business.[21]

The aeronautical committees of Kent and Cuyahoga Falls Chambers of Commerce and the Stow Civic Association also saw new opportunities for Stow Field with the defeat of the Akron bond issue. Meeting in December 1927, they discussed the airfield as a possible site for hangars to house Goodyear's proposed super-dirigibles, as well as for a public airport. The cost would be borne by Goodyear. The group soon came together again upon learning that Continental Air Lines had proposed Stow Field specifically (instead of Akron in general) as a stop on its Cleveland-to-Louisville airmail route if the necessary accommodations were in place. This shifted the discussion to improving the field only for heavier-than-air machines and not for zeppelins.[22]

The committees met in January to discuss the possibility of a $100,000 bond issue divided among the three localities to support the development of Stow Field as a community airport. The civic leaders in Kent reportedly scoffed at the probability of passing such a bond issue. Robbins countered with a request to raise $7,000 to install lighting to induce Continental Air Lines to use Stow Field for night stops on its airmail route. The US Post Office had established lighted airways across the country with a system of rotating beacons and course lights and initiated regular night airmail service in July 1924. For an airport to become an airmail stop, support for night flying was required. The airmail contract was particularly prized by the airline because it would support its developing passenger service.[23]

In May, Robbins came back to the tri-city group on behalf of Akron Airport, Inc., reporting that Stout Air Lines and Continental Air Lines were anxious to make Stow Field a regular stop on their routes. Funding was still needed, but his request that the three cities invest $6,000 for field improvements was turned down. While improvements at Stow Field were stalled, Robbins's Waco business prospered with a reported fourteen biplanes sold in the first part of 1928.[24]

Akron City Council decided early in 1928 that Fulton Field should become the site for Akron's municipal airport. The council wanted to demonstrate to Goodyear that the city could provide support for the dirigible plant since there was now serious competition from Cleveland, Baltimore, Los Angeles, and San Diego. Cleveland had just received news that the Glenn L. Martin plant was leaving for Baltimore, adding to its urgency to make a bid. Akron City Council elected to use funds from sale of a city building to acquire Fulton Field and adjoining properties.[25]

Akron moved ahead with improving Fulton Field for an airport by leveling and installing lights in preparation for airmail service. The field would be a night stop on the Cleveland-to-Louisville airmail route, which was scheduled to begin on July 1.[26] On October 27, 1928, the president of Goodyear-Zeppelin, Paul W. Litchfield, officially announced that the dirigible hangar would be built at Fulton Field. Costing between two and three million dollars, it would be the largest building in the world.[27]

Now that Akron Municipal Airport was becoming a reality, Robbins saw a growth opportunity for his business. Robbins Flying Service was granted permission by the Akron Airport Commission to erect a temporary hangar at the municipal field. When the city began preparing the land for the building of the hangars in May 1929, Robbins was already providing passenger flights and sales demonstration trips from the field.[28]

Aviation College

Realizing that Akron was now the focus of aviation activity, Robbins expanded into that market in 1928 by creating Aviation College. The Akron location promised "a practical education required in every branch of Aeronautics." By May 1929, enrollment had grown to nearly two hundred students. In addition to classroom instruction, courses included airplane rigging and construction, repair and assembly, and engine troubleshooting. A welding course was added to prepare students for working on the manufacture of airplanes. Students wishing to finish their ground school work and learn to fly would continue on to Stow Field.[29]

Frankie G. Renner

Robbins had assembled a strong staff at Stow Field that met all license requirements. With promising business prospects in the summer of 1929, Robbins placed classified ads in a trade publication for a qualified transport pilot and for two licensed airplane mechanics.[30] The employee who would be with Robbins the longest and whose tenure at the airport would exceed his own was Frankie G. Renner. She was first his student, and he later hired

Aviation College students working on airframes and propeller, Akron, 1929 (Courtesy of the International Women's Air & Space Museum, Cleveland, Ohio)

Frankie G. Renner standing next to a Robbins Flying Service Waco biplane at Stow Field, 1929 (Collection of Rick & Karolyn Gardner)

Frankie Renner's Federation Aeronautique Internationale sporting license, autographed by movie stunt flier Dick Grace, 1930 (Courtesy of the International Women's Air & Space Museum, Cleveland, Ohio)

To Frankie — Always be my friend.

Federation Aeronautique Internationale

Dick Grace

UNITED STATES OF AMERICA

ANNUAL SPORTING LICENSE ISSUED TO

M Frankie G. Renner

Place of Birth Sugar Creek, O.

Date of Birth July 27, 1909.

Orville Wright
Chairman
N. A. A. Contest Committee

1930

License No. 314.

F. A. I. Certificate No. 7410

Issued by N.A.A.

Type of Aircraft Airplane.

Frankie G Renner
Signature of Licensee

Valid until December 31, 1930

Frankie G. Renner (Collection of Rick & Karolyn Gardner)

Frankie G. Renner (1897–1985) began her career in 1919 as a stenographer in Akron. Her fascination with flying led her to use the small inheritance from her father for flying lessons from Hugh C. Robbins at Stow Field. Impressed with her dedication to flying and her background in office work, Robbins hired her as secretary and treasurer for Robbins Flying Service. Renner became the first woman in Akron to get a private pilot license, followed by earning a limited commercial license and a transport license. Renner instructed students, assisted in mechanics classes, and helped with the demonstration and sale of Waco airplanes. She also regularly performed aerobatics at Stow Field and other aviation events in the area. Renner made national news in March 1931 with her attempt to set the women's altitude record, falling short by a little more than three thousand feet. In 1932, Renner and twelve other local aviatrixes formed the "Squadron of Death" to share their interest in aviation and dispel superstitions about flying. Renner experienced two hangar fires at Stow Field—in 1929 and 1933–the second ending both the business and her aviation career. She went to work for Allen T. Simmons, owner of radio station WADC and a financial backer for Stow Field. She produced one of the first radio programs for women and was inducted into the local broadcasters' hall of fame. Recognized as an aviation role model for women, Renner died in 1985 at eighty-seven.[31]

her as secretary and treasurer for his business. In October 1929, she became the first Akron woman to obtain a pilot license. The Department of Commerce reported that at that time only 111 women in the United States were licensed to fly. Four of those were in Ohio, and the two with private licenses were Renner and Isabelle Chappell of Akron. By November, Renner was doing formation flying, including for the ring-laying ceremony at Akron Municipal Airport (hereafter referred to as Akron Airport) for the large dirigible under construction. She also helped with the Waco sales business. In the winter of 1929, she closed a contract for five Waco biplanes and also assisted United States Air Lines in Cleveland in becoming a Waco distributor. The Federation Aeronautique Internationale granted Renner lifetime membership and issued her a sporting license to participate in their competitions. The Federation gave official recognition to pilots for records and sanctioned racing contests.[32]

Under Robbins's management, aerial exhibitions continued to be offered at Stow Field, but without the intense spectator interest of the early years. One of the most publicized was in November 1928 when Freddie Lund arrived in his Waco biplane promising a thrilling one-man air circus. Lund was the chief test pilot for Waco Aircraft Company and was known for being the third man to successfully do the "outside loop" maneuver, first accomplished by Lt. Jimmy Doolittle. Although the outside loop was barred from Stow Field, two thousand spectators still showed up to see Lund perform military stunts in his commercial airplane.[33]

Hangar Fire at Stow Field

Catastrophe struck Stow Field on March 6, 1929, when a fire of unknown origin destroyed the hangar, adjacent office, and eight airplanes. Five of the airplanes destroyed were privately owned. Three airplanes were saved along with the motor of a fourth. Also destroyed in the fire were lockers with flying suits and parachutes and spare parts. Secretary Renner saved the company's records. Robbins and chief pilot F. A. Alden Jr. were slightly burned trying to retrieve airplanes as high winds whipped flames through the wooden frame hangar. The loss was estimated at $50,000. Fred Smith, as owner of the field and hangar, reported that insurance had been allowed to lapse and that the ten-year-old hangar was worth $5,000. Robbins announced that efforts would begin immediately to replace the hangar with a steel one and that all other activities would resume.[34]

Flames sweep airplane hangar at Stow Field, March 6, 1929 (Courtesy of the *Akron Beacon Journal*)

Robbins reported two weeks later that a contract had been awarded to build a new steel-covered 50 x 60 ft. hangar, with an adjacent workroom and locker room. When completed, he planned to build a second hangar and office building. The buildings would be located on the west side of the field, opposite from the former hangar location. Robbins began ordering replacement Waco biplanes, and by the end of July there would be a fleet of ten.[35]

Competing Airports

In addition to the losses suffered from the hangar fire, Robbins's business at both Aviation College and Stow Field now faced significant competition. The new players were Air Services, Inc., in Akron and Issoudun Mid-City Airport in Darrowville.

The new business in Akron had the underwriting of Akron industrial leaders. One of their first projects, announced in January 1929, was to support a unique aviation school that would offer engineering courses in addition to the usual ground school and flying instruction. The courses would be taught at the accredited engineering program at the Municipal University of Akron, generally referred to as Akron University. This, along with Akron's lighter-than-air accomplishments, would make Akron "an aeronautic mecca."[36]

Air Services, Inc., would manage Akron's aviation school. Coursework would be at the university and flight instruction at the municipal airport. Classes began in June 1929.[37] Although the financial support Air Services received was in sharp contrast to the modest investment Robbins had when he started Aviation College, there is no evidence that Air Services' school was more successful.[38]

After announcing Akron's new aviation school, the project's supporters sought unsuccessfully to have Charles Lindbergh visit and review the plans. However, Amelia Earhart did accept an invitation to go over the proposed plans for women fliers enrolled at the school and made her visit in February 1929. She would make another trip to Akron in August 1931 to help celebrate the christening of the airship USS *Akron.*[39]

Meanwhile, five miles north of Stow Field, a new airport was under construction in Darrowville. When it was no longer under consideration for Akron's airport, private investors moved to lease a thousand acres for a commercial airport. One story reported that Lindbergh had identified the site as a possible midwestern airbase for a transcontinental airline and that he worked with his friends Maj. Thomas G. Lanphier and Col. Henry S. Breckinridge in establishing the airfield.[40]

The business was incorporated as Issoudun Mid-City Airport. The principals included Lanphier, president; Carl K. Wollam, vice president and general manager; and Breckinridge, chairman of the board. The name "Issoudun" was chosen because Lanphier had served as the commander of the famed World War I flying school at Issoudun, France. Issoudun Mid-City Airport would be a memorial to the airmen who had trained at Issoudun and would continue its training traditions. "Mid-City" referred to the location being midway between Akron and Cleveland. Wollam was a well-known Akron balloonist and had been a crew member on the *Pilgrim* on the day of its christening. Breckinridge served as assistant secretary of war under President Woodrow Wilson and was an advisor to Lindbergh.[41]

The plan called for a full-service airport with a flight school, two landing fields (one for students, the other for commercial purposes), a hangar for thirty airplanes, airplane service, lighting for night flying, and a nine-pump service station. Construction of one runway necessitated moving seven thousand square yards of dirt. When fully developed, the company expected the airport to be "the largest east of the Mississippi River." The school would offer different types of pilot training, costing from $300 to $3,000. The enrollment target was two hundred students, and a fourteen-room brick house was remodeled to serve as a dormitory. The first class in September 1929 drew twenty students from Cleveland and ten from Akron.[42] After opening in August 1929, Mid-City began to be used as an emergency landing field for passenger airplanes traveling the Washington-Pittsburgh-Cleveland route and airmail airplanes flying from Pittsburgh to Cleveland.[43]

On December 27, Charles Lindbergh and his wife, Anne, made an unannounced stop at Mid-City, arriving from Curtiss Field in Valley Stream, New York. The stop allowed Lindbergh to see the airfield his friends Lanphier

Issoudun Mid-City Airport, located in Darrowville, Ohio, between Hudson and Stow, 1929. "Issoudun" was later dropped from the airport's name. (Courtesy of Hudson Library and Historical Society)

and Breckinridge were developing, as well as to refuel. The newspaper account found some irony in their unexpected visit since a half-dozen officially announced visits by him to the area over the years never materialized. It is believed that Lindbergh made frequent secret visits at Mid-City in its early years, but this December 1929 stop is the only known one of the famed aviator in the Akron area. Despite many stories to the contrary, there is no indication that he ever visited Stow Field.[44]

Ohio Aviation Legislation

The Ohio legislature in 1929 passed strong legislation to regulate intrastate flying similar to the rules that the Air Commerce Act of 1926 had imposed on interstate and foreign transport. The Ohio law required that every pilot be tested and licensed and every airplane inspected for airworthiness, thereby putting an end to the itinerant or "gypsy" fliers operating in Ohio. While this step to improve air safety had merit, what was really needed were uniform regulations for all states because state laws varied or were nonexistent. It was not until December 1, 1941, that all pilots and aircraft in the United States were required to have federal certification.[45]

With this law, Ohio became one of the first states to create a state aeronautical department and to make "air marking" compulsory for all incorporated places. Air marking was the placement of town names where pilots could see them from the air to help guide them to an airport. Pilots frequently found themselves uncertain of their location and unable to find their intended landing field. Mapping was slowly improving, but navigational instruments were limited. One story had Lindbergh missing Steubenville and inadvertently doing his planned flyover at Youngstown instead. A technique pilots used was called "shooting the depot," which referred to flying low to see a town's name on a railway station to get needed bearings. Unknown obstructions made this hazardous. Railroad tracks were referred to as the "iron compass," as they provided a reliable navigational guide.[46]

Ohio's air marking law became effective on June 25, 1929, but some steps had already been taken to address the problem. After losing his way while flying, Will Rogers made an offer to Ohio civic groups to pay for the paint needed to mark their town names in large letters on buildings. Given the strong response, Rogers soon modified the offer by limiting it to cities with only four letters in their name. The Kent Chamber of Commerce said they would relieve Rogers of furnishing paint and handle it themselves. "Kent" was painted on the roof of Robert "Bob" Redmond's Studebaker garage with an arrow pointing to Stow Field.[47]

Sherwin-Williams Company stepped in and offered to provide paint for the first hundred Ohio cities. The letters were to be 15 x 10 ft. and painted white on a yellow background. A Cleveland outdoor advertising firm, Harry H. Packer Company, joined the effort by volunteering to do the roof painting, and the *Cleveland Plain Dealer* served as the clearinghouse for requests. A newspaper account a year later said that Kent was marked with four signs on each side of town that pointed to Stow Field and gave compass directions.[48]

Future Notable Pilots at Stow Field

Stow Field introduced thousands of local residents to flight over the 1920s, whether through aerial exhibitions or airplane rides. For some it was the start of a lifelong career in aviation.

William Marsh "Bill" Bower of Ravenna rode in an open-cockpit biplane at the age of eight when his grandmother took him to Stow Field in 1925. An aviator in the US Army Air Corps, he joined the 17th Bombardment Group in 1941. In 1942, he volunteered to fly one of the sixteen B-25B Mitchell bombers for the first American aerial attack on Japan, known as the Doolittle Raid, led by Lt. Col. James "Jimmy" Doolittle. For his role, Bower earned the Distinguished Flying Cross. Colonel Bower died in 2011 as one of the last of the surviving pilots of the Doolittle Raiders.[49]

William Grafton Draper of Cuyahoga Falls recalled that at age seven, he and a pal would hike along the train tracks to Stow Field, where he would regularly be thrown out of the hangar. His first ride was in 1927 in a Robbins's Waco biplane. Draper was a transport pilot in the US Army Air Corps in 1942 and was selected in 1950 as the personal pilot for US Gen. Dwight D. Eisenhower, supreme commander of the Allied Forces in Europe. He continued as Eisenhower's pilot during his presidency, suggesting "Air Force One" as the designation for the president's airplane. A heart attack ended Colonel Draper's career, and he died in 1964 at age forty-four.[50]

Twins Joseph C. and Howard C. Funk were born in Akron and took their first flight in the late 1920s in an open-cockpit biplane at Stow Field. They learned to fly gliders at Akron Municipal Airport by using the updrafts of the Goodyear Airdock and went on to build their own glider. Realizing that the glider's safety features could be built into an airplane, they designed an airplane with devices to prevent stalling and spinning. When their Akron manufacturing company had to liquidate in 1941, a Kansas firm continued manufacture with the participation of the Funk brothers. It is believed that about two hundred Funk airplanes still fly.[51]

Summary

There was great promise in 1927 when Hugh C. Robbins took over operation of Stow Field with Robbins Flying Service. An experienced pilot, Robbins soon demonstrated he was also an enterprising businessman willing to take risks. He established himself in Akron with Aviation College and repeatedly sought funding to improve and grow the business at Stow Field. The future for aviation was promising following Charles Lindbergh's flight and the newly enacted federal and state legislation. Airmail service continued as the motivation for the establishment of airway routes and municipal airports. But as 1929 ended, Robbins was rebounding from the devastating hangar fire at Stow Field and facing new competition from airports in Akron and Darrowville. Those challenges would pale in comparison with those presented by the oncoming Depression.

Chapter 4

The Depression Sets In, 1930–38

In mid-1929, US secretary of commerce Robert P. Lamont stated that "nothing stands in the way of the full development of air transportation in all its phases and all its possibilities." Ample capital was available for development. That would change when aviation stocks and securities were among those hit in the stock market crash of fall 1929. Upon reflection in November 1930, Clarence M. Young, assistant secretary of commerce for aeronautics, stated that the aviation industry as a whole had been overpromoted in 1928 with outsized expectations.[1]

Effects of the Depression

As the federal government worked to "readjust its financial position," the Bureau of Air Commerce managed with a reduced budget to maintain its regulatory functions and development of the national airways system. Mileage in the system would still grow 66 percent from 1930 through 1937, with over 97 percent of the routes lighted. The most resilient sector of aviation even in the early years of the Depression was air transport, which was sustained by revenue from airmail contracts. Airmail traffic grew from nearly 20 million miles flown in 1930 to over 36 million in 1932. Provisions in the Air Mail Act of 1930, known as the Watres Act, changed the payment structure for transporting airmail, encouraged the use of larger aircraft, and stimulated passenger service. The result was that scheduled passenger service, which was in its infancy in 1928, reported that 84 million passenger miles were flown in 1930. That increased by 51 percent to more than 127 million passenger miles two years later.[2]

Other areas of aviation, however, were very hard hit. Aircraft manufacturers suffered with sharp declines in production from 5,400 units in 1929 to only 900 in 1932. General aviation experienced serious reductions in business activity as the Depression devastated the customer base for flight training, passenger services, and airplane sales. From 1929 to 1932, miles flown decreased by 29 percent and passengers carried for hire or pleasure by 43 percent.[3] Now, even fewer students could afford to take ground school courses or go on to do flight training. Airplane sales were similarly stressed. The period of 1930 through 1938 proved to be a survival challenge for Stow Field, as well as for Akron Municipal and Mid-City Airports.

In the Akron area prior to the economic downturn, a worker's typical schedule was an eight-hour day and a forty-eight-hour week. By October 1930, Goodyear, for example, had reduced workers' hours to avoid layoffs by instituting a twenty-four-hour week with six-hour shifts, four days a week. To help the unemployed, Akron created an apple selling program and turned vacant lots into "opportunity gardens." Hard times continued in Summit County in 1932 as Akron and other municipalities faced a financial crisis because of a "collapse of tax collection." All public payrolls and services were affected, and the mayor of Akron proposed reducing all payrolls beginning in May by 50 percent.[4]

Unemployment in the aviation industry led to the announcement in 1932 by the National Air Pilots Association that it was disbanding due to the inability of members to pay dues. Both Hugh C. Robbins and his secretary, Frankie G. Renner, were members. Robbins had two businesses to keep afloat—Robbins Flying Service at Stow Field and Aviation College in Akron. He actively worked to attract students to Aviation College who might then go on to take flight instruction at Stow Field. Promotion included heavy use of classified advertisements during the first half of 1930. One offered a "free course—financed by Akron men and capital." Another highlighted a special aircraft welding course for $125 and that "terms may be arranged," followed later by notice of an airplane mechanics class for ladies.[5] The tagline of being "Akron's own college" was used in the classified advertisements over several months—a strong statement, given that Air Services' school across town was a collaborative effort with Akron University and had underwriting from Akron business leaders.

Summit Flying Club

In January 1930, students at Aviation College formed the Summit Flying Club. Flying clubs provided a means for members to reduce the cost of flying by combining their resources to purchase an airplane and hiring an instructor or a qualified member to teach them to fly. Airplanes had to be certified, and instructors needed to have a transport license. The National Aeronautic Association promoted the movement as a means to draw more individuals into flying. Summit Flying Club members purchased an airplane and readied it for flight as part of their coursework at Aviation College. The airplane was then housed at Stow Field, where Robbins provided instruction, cutting the cost of flight training by two-thirds.[6]

By 1931, the Depression was taking hold, and all activity at Aviation College came to a virtual halt. The last notice was a newspaper advertisement in April for an introductory aviation course that included a free airplane ride for only fifty dollars.[7]

Activity at Stow Field in 1930

Gliders also gained interest nationally in 1929. The advantages put forward by enthusiasts were that they offered a cheaper and safer way to fly. The safety argument was disputed by others, as well as the suggestion that a course in gliding should be a prerequisite to flight training.[8]

Locally, the *Akron Beacon Journal* ran a series for youth on how to build a glider model, complete with diagrams and instructions. In April 1929, members of the Akron Gliding Club brought their training glider for testing to Stow Field. Purchased from Heraclio Alfaro, a Cleveland aeronautical engineer, the glider was piloted by Dr. Wolfgang Klemperer, a famed glider pilot and design engineer for Goodyear-Zeppelin. A ground crew of a dozen men launched the glider with its twenty-foot wingspan from the knoll at Stow Field using a rubberized rope. The glider's longest flight lasted seventeen seconds. Glider enthusiasm infected the members of the Summit Flying Club, too, leading them to obtain a new glider the next month, which they assembled and tested at Stow Field.[9]

For Stow Field, 1930 was relatively quiet. Some business that previously would have come to Stow Field now went to the new Akron Airport. Robbins observed, "lighter-than-air interest is paramount in Akron," overshadowing interest in airplane travel. He continued to promote his Waco business and in April attended the Detroit All-American Aircraft Exposition. He

Above: Reverend Smith's eighty-fifth birthday ride in a Robbins's biplane, 1927 (Fran M. Hermance Collection)

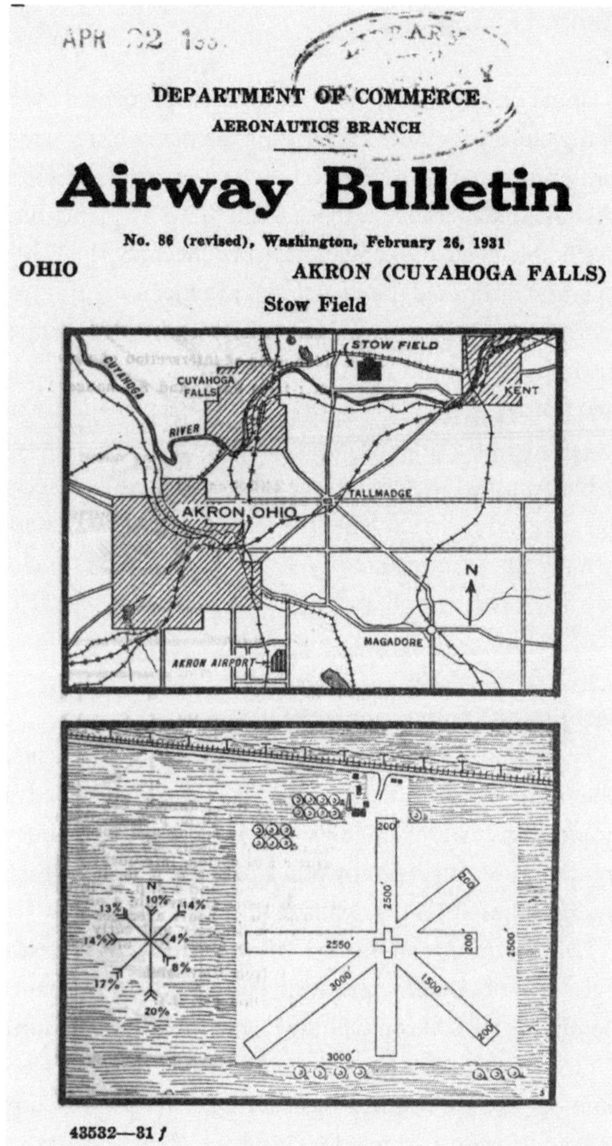

DEPARTMENT OF COMMERCE

AERONAUTICS BRANCH

Airway Bulletin

No. 86 (revised), Washington, February 26, 1931

OHIO **AKRON (CUYAHOGA FALLS)**

Stow Field

43532—31

Right: "Ohio-Akron (Cuyahoga Falls)-Stow Field," *Airway Bulletin,* no. 86 (rev. Feb. 26, 1931). The hangar is now located on west side of field. (Courtesy of HathiTrust, https://hdl.handle.net/2027/uc1.b3122366?urlappend=%3Bseq=111)

came back with Waco's new model F—a three-place open-cockpit biplane, smaller and lighter than other models. There would, however, be no news of airplane sales that year. A revised *Airway Bulletin* for Stow Field issued in 1931 clearly showed the runways along with Robbins's new hangar on the west side of the north-south runway.[10]

The owner of Stow Field, Rev. Edward J. Smith, died on April 16, 1930, at age eighty-seven, and his son Fred became the owner. A forty-year resident of Stow Township, Reverend Smith lived on the property with his son where he could watch airplanes from his window. Their house was down the lane from the big house he had built decades earlier. He had witnessed the history

Nick Firestone, Frankie Renner, Dick Grace, and Hugh Robbins at Stow Field, 1930. Grace encouraged Renner's flying and was a friend to Robbins. (Courtesy of the International Women's Air & Space Museum, Cleveland, Ohio)

of Stow Field play out from its earliest days and became known as the "flying minister" for his rides over the surrounding area.[11]

A personality who was a presence at Stow Field in the early 1930s was Dick Grace, a movie stunt flier for such movies as *Wings, Lilac Time,* and *Air Circus.* A longtime friend of Robbins, he worked on a story about movie stunt fliers during a visit to Stow in summer 1930. Called "The Lost Squadron," the story was turned into a movie starring Richard Dix and Mary Astor. Grace tried to enlist Robbins to come to Hollywood with him to fly in the movie, but Robbins declined. Like most of Grace's more than thirty intentional crashes, his role in this movie was uncredited. During his area visit, Grace was also seen at Halle's Department Store in Cleveland signing copies of his book called the *Squadron of Death.* The book followed seven movie stunt fliers who belonged to his club with that name. Grace was the last survivor. When speaking at Aviation College in September, he underscored the need "to be more particular about little details of a flight than about the major things."[12]

When the Affiliated Exchange Clubs of Ohio held their convention at Ashtabula Municipal Airport in June 1930, seven airplanes flew from Stow Field, including those piloted by Robbins, Renner, and Grace. Grace did stunt flying with Robbins, who also stayed on to do night flying with his specially illuminated airplane. That summer, Renner continued to establish her credentials as a female aviator, becoming the second woman in Akron and the third in Ohio to obtain a limited commercial license. She, too, was seen doing stunts at local fairs, such as balloon bursting at the Aerial Carnival at Cleveland Airport.[13]

Frankie G. Renner's Altitude Attempt

Big news came at the start of 1931 when Stow Field announced that Frankie Renner would attempt to break the women's altitude record of 27,418 feet, which was held by Elinor Smith of Freeport, New York. Renner and Hugh Robbins worked with engineers at the Waco Aircraft Company, as well as with the chief engineer of the Wright Aeronautical Corporation, to prepare the airplane that had been loaned to Robbins Flying Service for Renner's attempt. The Waco CRG biplane (NC600Y) was one of two built in 1930 to compete in the 1930 Ford Reliability Tour. The airplane was a straight-wing Waco with a J-6, seven-cylinder Wright Whirlwind motor predicted to be capable of operating above 30,000 feet. Renner had begun doing experiments for Waco the previous year to obtain data that would aid in the design of military airplanes for high altitudes.[14]

Renner's airplane had a supercharger along with several other unique features. The open cockpit was lined with heavy felt for extra insulation, enclosing her up to the neck "like a giant sleeping bag." She supervised the development of a novel electrical system of individually controlled heating pads for her flying suit, moccasins, and gloves. Her goggles were also heated to prevent her eyeballs from freezing, as had happened to Maj. R. W. Schroeder in his record-setting altitude flight in 1920.[15]

Frankie G. Renner in her gear for her attempted altitude record flight, March 1931. The Waco CRG biplane (NC600Y, Construction Number 3349) was one of only two built specifically for the 1930 Ford Reliability Tour. (Courtesy of the International Women's Air & Space Museum, Cleveland, Ohio)

Although February usually offered the most favorable atmospheric conditions to achieve an altitude record, Renner had to wait for suitable weather. During her wait, Renner made several test flights that allowed her to adjust the propeller setting and uncover a faulty connection to her liquid oxygen supply. On the test flight of March 5, she reached a height of 23,000 feet. Finally, her flight aiming for a record of 35,000 feet was set for Friday, March 13. The weather report indicated some clouds at the onset and then good conditions at higher altitudes. Sixty-three gallons of fuel were loaded into the airplane, and it was washed with an alcohol-glycerin mixture to prevent freezing.[16]

In Renner's handwritten account of the flight, she shared that she wore two pairs of wool stockings, three pairs of booties, heavy moccasins, two sweaters, three pairs of heavy fleece underclothing, two wool scarves, a fur-lined winter flying suit, oxygen helmet, gloves, and goggles. She forewent taking a parachute to lighten the load. Once dressed, she was driven from the Stow Field office to her airplane to avoid picking up any ground dampness.[17]

The flight was expected to take a little more than three hours, most of which would be spent climbing. The record attempt was under the auspices of the National Aeronautic Association, with official government supervision from Lt. T. G. W. Settle of the navy inspection bureau. Official observers were B. E. "Shorty" Fulton, of Akron Airport, and H. W. Maxson, of Goodrich public relations. Robbins was the flight's sponsor.[18]

Renner's flight began at 1:42 P.M. and lasted for four hours and four minutes. Renner noted that at 3 P.M., she was lost for a period, having flown into dense clouds that obscured earth and sun. She reported her altimeter stopped recording at 3:23 P.M., showing 28,000 feet. She continued climbing until her fuel indicator was at zero and then began her thirty-one-minute descent. She estimated there were winds of 40 mph and temperatures of forty degrees below zero, producing blood in her mouth and frosting over her goggles. The Department of Commerce physician examined her upon landing and found her condition to be excellent except for frostbitten feet. Lieutenant Settle sent her barograph to Washington for evaluation by the National Aeronautic Association.[19]

Renner received word in late March that her flight failed to set a record as her calibrated barograph recorded a height of only 24,990 feet. She planned to try again within several weeks. Her airplane was being fitted with a 300-horsepower motor and specially built supercharger and carburetor. She said she would carry two barographs with her instead of just one. The flight was postponed until October. By January 1932, things had changed, and she was quoted as saying, "No, I am not going after the altitude record again. I'm getting so I take my flying straight."[20]

The End of Robbins Flying Service

Robbins's efforts continued to bring in spectators to Stow Field in the summer of 1931, despite competition from Akron Airport. Although a crowd of several hundred people showed up in July to see stunt flying and parachute jumping, times were increasingly difficult. The Stow Field canteen with its gas station and barbecue went up for sale in October 1931.[21]

All was quiet at Stow Field going into 1932. Robbins's flight school reported several new students. Robbins was active as a Waco representative, and by April additional Ohio counties were added to his territory, bringing the total to fifty. He announced in June that his advertising had paid off with a long-distance sale to a flight club in the Panama Canal Zone.[22]

That airplane sale was the only good news, as later in June 1932, Akron Airport, Inc., Robbins's company held with Allen T. Simmons, was in receivership. This ended the very active and promising period of Robbins as operator of Stow Field. The tipping point may have been the inability of customers to pay for new Waco airplanes that Robbins had ordered for them. Fulton was one of those appointed by the court to serve as an appraiser for the company's assets. Fred Smith, the owner of Stow Field, indicated he would seek new tenants.[23]

After several weeks, an application for a new corporation for the airfield was approved by the Ohio secretary of state, authorizing the issuance of 250

Twin Coach, a Kent bus manufacturer, uses its Model 30 and Model 40 busses to spell out *TWIN* at Stow Field, in the early 1930s. The aerial photo shows the replacement hangar Robbins built in 1929, which would be destroyed in a fire in 1933. A gas station is visible near the airport entrance. (William D. Schloman Collection)

shares of stock. The new corporate name was Stow Flying Field. Frankie Renner served as manager, becoming only the second female in Ohio to assume this role. Akron School of Aeronautics moved to Stow Field to provide flight training. August brought another business loss at the field with a classified ad offering the sale of the "gas and lunch room" at Stow Flying Field. The owner would sacrifice for a small down payment.[24]

Squadron of Death

The biggest news for area aviation came in spring 1932, when Renner and eleven other area aviatrixes formed an organization to encourage an interest in flying by women. They chose the name "Squadron of Death," the same

Squadron of Death photo with Frankie G. Renner in the center of the bottom row, 1932 (Courtesy of the International Women's Air & Space Museum, Cleveland, Ohio)

used by Dick Grace for his squad of movie stunt fliers, and sought Grace's permission to use the name. Renner was elected president, and Isabelle Chappell was chosen as secretary and treasurer. The group's insignia was a skull and crossbones with the initials S.O.D., and a black crow was chosen as its mascot. The members soon selected another member, bringing their number to thirteen. An account many years later discounted the claim that members wore high heels when flying.[25]

It is not clear why, but in June members changed the group's name to "S.O.D. Fliers." They also chose to honor Mildred Stinaff, a young woman who had died while stunting the previous year. Her death came before their group formed, but they expected that she would have joined them. Renner flew over Stinaff's grave at Standing Rock Cemetery in Kent, dropping American Beauty roses.[26]

Second Stow Field Hangar Fire

Once again a financial backer of the airfield, Allen Simmons made an announcement in January 1933 that activities at Stow Flying Field were temporarily suspended until weather conditions improved in the spring. Some operations resumed in April, and business prospects were promising. Renner now had her transport license. A high-wing monoplane was being constructed on the premises, and two glider clubs planned to use the field for their activities. However, the message in the *Air Commerce Bulletin* of May 15, 1933, seemed to indicate the suspension was more than just a weather delay, saying the field "is no longer available for the use of aircraft."[27]

The death knell for Stow Flying Field came on August 14 with a fire that began in the early morning hours, destroying the hangar and doing an estimated $10,000 damage. When Hugh Robbins built the hangar in 1929, it was constructed with a corrugated iron exterior and a wooden interior. The fire destroyed three airplanes, two of which belonged to Simmons; a glider; and spare parts. Some chickens in a nearby coop were all that were reported saved. The cause of the fire was undetermined. Simmons reported the loss was partially covered by insurance and that he expected to rebuild, but that did not happen. Within four days following the fire, Frankie Renner was hired by Simmons to work at his radio station WADC and would never fly again.[28]

Renner's relatively short aviation career had defied society's expectations for women, which still rigidly defined women's roles despite the victory for suffrage in 1920. Renner, like the other women pilots of this period, found flying liberating. It conveyed a new sense of mastery and confidence. Renner was at the forefront of women locally and within Ohio by obtaining her various pilot ratings. Her record-setting attempt, although unsuccessful, was still considered a remarkable feat and showed what women could do. Demonstrating competence as a female pilot promoted aviation as safe and achievable and contributed to her effectiveness as a Waco salesperson. Renner and her colleagues in the Squadron of Death supported each other and modeled for other women the feasibility of becoming a pilot, or, one imagines, to take on challenges they had not thought possible. Renner's final aviation-related position as manager of an airfield was also atypical for a woman at the time. Once the hangar fire ended that business, Renner's chance to make a living in aviation ended as well.[29]

After the second hangar fire, Stow Field lost its status as a commercial airport because it was without proper management or facilities. Aeronautical charts no longer showed Stow Field, but it would reappear in 1935 marked as an auxiliary field. Auxiliary fields did not provide services for airplanes, nor were they used as ports for passenger or cargo operations. The owner of an auxiliary field would designate it as such, making it available for airplanes to land and take off without charge. The pilot's use of an auxiliary field was entirely at his own risk, as the field's owner made no guarantee of conditions.[30]

Postfire Field Management

There are no reports about the airport property until notice of a sheriff's sale ordered in May 1936. A sum of $12,500 was owed to the Superannuates Fund Association of the Northeast Ohio Conference of the Methodist Episcopal Church, for an apparent mortgage that Reverend Smith had obtained through the Methodist church. In an arrangement with the court, his son Fred was granted one acre with the small house where the two had made their home in recent years.[31]

A newspaper article reported in March 1937 that the airport was to be reopened with new management and that plans included construction of a new hangar. It had seen little use as an auxiliary field for over three years, so it must have been in rough shape. Flying did resume in April with three airplanes—one closed-cockpit and two open-cockpit models. Flight instruction and commercial flying were offered, returning Stow Field to the status of commercial airport. Art Laurence was the chief flight instructor and would provide stability for the next several years.[32]

Significant federal legislation was passed with the Civil Aeronautics Act of 1938, which created the independent Civil Aeronautics Authority (CAA)

Above: Art Laurence, chief flight instructor, on the left in front of new hangar at Stow Field, 1937 (Garrett Family Collection)

Top right: Marshall Garrett in airplane at Stow Field with new hangar in background, 1938 (Garrett Family Collection)

Center right: Flight student Leo Schneider in Great Lakes 2T-1A Sport Trainer, Stow Field, 1939 (Leo and Betty Schneider Collection)

Bottom right: Fliers at Stow Field, 1938–39. *Left to right:* James Woodall (airport manager), Chester Piotroski, Marshall Garrett, Art Laurence. (Garrett Family Collection)

along with an Air Safety Board. The new act focused on aviation safety and economic regulation of civil aviation, shifting authority from states to the federal government regarding such issues as the airworthiness of aircraft, competence of airmen, and the certification of both aircraft and airmen. Now, more broadly defined airmen would be required to pass a competency examination to receive certification.[33]

The first student to receive his pilot license at Stow Field under the new civil air regulations was Marshall W. Garrett, whose career decades later would lead to being manager of this very airport under Kent State University ownership.[34] Another area student was Leo A. Schneider, a young man in his early twenties from Ravenna pursuing his personal interest in flying. With flying having resumed, Stow Field appeared to be at least tentatively back in operation.

Akron Airport during the Depression

Akron Airport experienced highs and lows during the Depression. The airport's new terminal opened with great fanfare in June 1931, followed by the inauguration of Pennsylvania Air Lines' airmail and passenger service. By far, the biggest event at the airport was on August 8, when Goodyear's dirigible the USS *Akron* (ZRS-4) was christened with crowd estimates of 50,000 to 100,000.[35]

By May 1933, the city of Akron faced insolvency. Industries in Akron were operating at 20 percent capacity. Sixty-two percent of homeowners were delinquent in paying their taxes, requiring a 50 percent reduction in the city's operating expenses. This halted the use of public funds for airport operations. The federal Civil Works Administration (CWA) program offered some relief by providing men for airport grounds work.[36]

Air Services, Inc., the major tenant at Akron Airport, was unable to pay rent and by spring 1935 was forced into reorganization. Over the next months, there would be a fight over hangar ownership, a hold on issuing a new lease, another reorganization plan, and a rejection of that plan by the city. The end came in April 1936, when Air Services went into receivership. A new firm called Akron Airways, Inc., was selected as the airport's new contracted operator. The two owners of the business were R. A. "Rudy" Van Devere, manager of the South Bend, Indiana, airport, and Chet Betz, former manager of Barberton's airport. Van Devere became sole owner of Akron Airways in July 1936.[37]

Van Devere would later tell the story of flying into Akron Airport one night and finding the hangar locked up. A man he mistook as "a drunken night watchman" turned out to be a sober Shorty Fulton, angry that the hangar was locked. Fulton shared with Van Devere that the current business was in default and that the city wanted a new operator. It was that first contact that resulted in Van Devere becoming the contracted operator at the airport and a key person in area aviation for the next decade.[38]

Rudy A. Van Devere (Van Devere Family Collection)

R. A. "Rudy" Van Devere (1905–1988) bought an airplane in 1928, got his license, and then established his first aviation business at Bendix field in South Bend, Indiana. In 1935, he learned of an opportunity to operate Akron's airport and landed the lease. He then founded Akron Airways, Inc., becoming a major player in area aviation for the next decade. Soon he expanded, leasing Mid-City Airport in Darrowville in 1939 and purchasing Stow Field a year later. Van Devere provided flight training for the Civilian Pilot Training Program at Kent State University and Akron University, using his facilities at Akron and Mid-City Airports. His school in Helena, Arkansas, trained over 3,000 air force cadets. In 1940, Van Devere qualified to join the Caterpillar Club for successfully using a parachute to bail out of a disabled aircraft after being dropped from an open-cockpit airplane during a slow roll. As the federal programs ended, Van Devere sold Stow Field minus its buildings to Kent State in 1942. The failure of private airplane sales to rebound after the war led him to seek new opportunities. Seeing growth in automobiles, he purchased a Packard auto dealership in 1947, followed by a Dodge-Plymouth dealership. He dissolved Akron Airways in 1953, the same year that he sold his airport buildings to Kent State. In 1954, Van Devere acquired an Oldsmobile dealership that has since grown and changed with the auto industry. Rudy Van Devere, one of the "major aviation contributors in the Akron-Canton Area," died on October 16, 1988.[39]

Over the next three years, the economy weakened further in what became known as the recession of 1937 through 1938. Akron Airways struggled and relied on offering airplane rides and flight instruction, to which Van Devere added a course in airplane mechanics to increase appeal.[40] The company hit a rough patch in August 1938, when it was four months delinquent on rent payments for the hangar at Akron Airport. Akron City Council canceled its lease and announced it was seeking bids for a new operator. Akron Airways turned out to be the sole bidder and was granted a new lease based on a bid with lower rental rates.[41]

Mid-City Airport Struggles

The area's third airport, Issoudun Mid-City Airport, struggled through the Depression with many management changes. Its training school had gotten off to a strong start in 1929, offering complete courses with new Brunner-

Winkle Bird airplanes. In 1930, the airport expanded, becoming a distributor for Aeronca airplanes. Excitement followed in July with news that discussions were under way for the airport to be the site of a stop on an air-rail route from New York to Chicago. Despite some suggestion of Charles Lindbergh's involvement with the plan, it did not go forward.[42]

By November 1931, the training school became the Aeronca School of Flying. Carl K. Wollam had taken over management of the airport and officially dropped "Issoudun" from its name. But the airport struggled, and Wollam terminated his Mid-City lease in 1932 and temporarily moved the Aeronca School of Flying to Akron Airport. A hoped-for merger with Air Services fell through, leaving the Aeronca school without an airport and Mid-City Airport without a tenant. Soon, Aeronca was back at Mid-City with Al Le Fevre as manager and chief instructor.[43]

For a short period in 1934, Mid-City was taken over to serve as an inland seaplane base. That failed venture brought Le Fevre back as airport manager later that year. By 1936, management had changed again to glider pilot Willis Sperry and Akron Airport's Ray Vaughn. By all reports, enrollment in their flight school in 1937 was good.[44]

The big news in January 1939 was that Rudy Van Devere had leased Mid-City Airport. This operation would be in addition to his business at Akron

Akron Airways, Inc., at the Akron Municipal Airport, 1937. Various training airplanes are on display outside the hangar along with Akron Airways' large Ford Tri-Motor on the left. (Van Devere Family Collection)

Kent State May Queen candidates on the wing of Akron Airways' Ford Tri-Motor, Stow Field, May 17, 1938. *Left to right:* Marge McNab, Bonita Fuscoe, Claudia Brooks, Dorothy Knapik, Peggy Maratta, Reita Mehlenbacher, Genevieve Sofranec, Betty Manchester, and Marie Smith. (William D. Schloman Collection)

Airport and would be managed by his younger brother Ray and Gail Mishler. There would be a flight school, storage, and airplane services.[45] The Van Deveres and Mid-City would play a key role for Kent State flying in the next years.

Summary

Stow Field, as well as Akron and Mid-City Airports, all struggled through the Depression. As the smallest operation, Stow Field was particularly hard hit due to its devastating fire and foreclosure. Akron Airport suffered once the excitement over dirigible construction passed but then had the good fortune to get Rudy Van Devere as operator with his eye for likely business ventures. The Mid-City lease changed hands many times but luckily also landed with Van Devere, promising stable operation in the years ahead. The struggle at Stow Field would continue, as would events that connected the three airports.

Chapter 5

Aviation Comes to Kent State University, 1939–41

As the United States emerged from the Depression, the aviation industry had yet to experience an economic recovery. Airports saw little to no new activity, and Stow Field was no different. However, over the next three years, federal investment in flight training would create a boom in the aviation industry that would firmly establish its prominence in the United States. While not every airport would survive the turbulent war years, those that became a part of the federal flight-training program were in a better position to flourish. Stow Field would find itself sitting on the sidelines for most of this activity. It would take new ownership to keep the airport from disappearing forever.

By 1938, with the outbreak of World War II still more than a year away, nations were busy preparing for a possible war. Aviation was bound to play a pivotal role in any future battle, making pilot training a necessity for most nations. In Nazi Germany, 65,000 young men were being trained as pilots and mechanics. Italy promoted aviation to students as young as eight years old through college and had a government-sponsored aeronautical society with 120,000 members. France had established 153 civilian-pilot training centers, where 20,000 young men between the ages of seventeen and twenty-one took primary flight training. The largest aviation educational effort was under way in Russia, where 600,000 students were studying aeronautics.[1]

In stark contrast, in 1938 the United States had no comparable aeronautical program. The US military had fewer than five thousand pilots and one thousand airplanes. The US Navy had one training base located in Pensacola, Florida, that could produce five hundred pilots a year. Likewise, the

US Army Air Corps had one training base, Randolph Field, located in San Antonio, Texas, that could turn out an additional five hundred pilots a year. Civilian flying across the nation was not very robust either. Records showed 21,118 certified pilots, but most had let their licenses lapse. Including 400 air transports, there were 9,732 aircraft in the United States.[2]

The Civil Aeronautics Act of 1938 had moved federal control over non-military aviation from the Bureau of Air Commerce to the newly created Civil Aeronautics Authority (CAA). Robert H. Hinckley, the new head of the CAA, envisioned a government-sponsored training program called the Civilian Pilot Training Program (CPTP). The program would partner universities and colleges with local fixed-base operators and flight schools certified by the CAA. The universities and colleges would provide the ground instruction, while the fixed-base operators and flight schools would provide the flight training.[3]

The private aviation industry economy was still suffering from the Depression, and Hinckley believed this new program could prime its recovery. Hinckley determined that "civilian lightplane aviation was on dead center if not in full retreat. Small fields were closing down by the score, others had such a small margin of profit that only love of flying kept their operators in business, lightplane builders were even closer to the brink of bankruptcy, and the entire industry was faced with disaster." In addition, the proposed program would provide a reserve of trained pilots should a wartime emergency arise, although the country was still adhering to a position of isolationism and neutrality.[4]

On December 27, 1938, President Franklin D. Roosevelt announced that he had approved the CAA plan to train college students. The program was unique among other government programs as it required no new departments or infrastructure because it would use established universities, colleges, airports, and flight schools. The Civilian Pilot Training Act was passed into law on June 27, 1939, with an appropriation of four million dollars for the first year of the program. This amount was less than half what was originally asked for, but it was still enough to train 9,885 new private pilots as well as refresher training for 1,925 instructors during the upcoming 1939–40 school year. While there were fewer than one hundred officially rated flight instructors in CAA records, anyone with a commercial or transport license could teach. Providing refresher training for those pilots who had let their licenses lapse would provide the needed flight instructors. The total cost came to less than $400 per person, including administrative costs.[5]

Civilian Pilot Training Program at Kent State University

The Ohio Bureau of Aeronautics, directed by Earle L. Johnson, assisted in bringing the Civilian Pilot Training Program to state institutions. In 1939, Kent State University was a thriving state university with an enrollment of 2,200. The CAA program came to the attention of the Kent State Board of Trustees, who voted unanimously at their August 1939 meeting to authorize KSU president Karl Leebrick to make a formal application for the CPTP.[6] On September 10, Kent State and seven other universities and colleges in Ohio were among the initial 166 schools to be accepted into the program. The final tally had three hundred schools accepted nationwide.[7]

The CPTP was open to both undergraduate and graduate students between eighteen and twenty-five years old who were US citizens and had an elementary knowledge of physics. One out of every ten students could be female. The CAA allotted thirty to fifty student slots to each school the first year. The CAA paid the school twenty dollars for each student. The students were selected based on scholarship, aptitude, and ability to pass the rigorous physical examination.[8]

Each college recommended an airport operator within the required ten miles of the institution to provide flight instruction. The CAA tried to limit payment to six dollars per hour of flight instruction, which was lower than the going rate. But since most schools had few or no students, the rate was attractive because a CPTP contract could save a school. Flight schools underwent a CAA inspection, and the CAA required one airplane with at least 50 horsepower for every ten students. The instructor had to be properly rated or rerated and standardized on CPTP teaching methods.[9]

The flight-training program was divided into three stages. Stage A was eight hours of dual instruction covering taxiing, taking off, and landing the airplane in different wind conditions; stalls and spins; straight and level flight; and turns of varying degrees of bank. After completing a minimum of eight hours of instruction and demonstrating proficiency, a student would be permitted to start Stage B and solo. Stage B started with the first solo flight, which was recommended to include three landings, followed by solo practice of Stage A maneuvers. Total flight time in Stage B was six hours.[10]

Stage C was the final stage and consisted of thirteen hours of solo flying and eight hours of dual instruction. A few new maneuvers were introduced and practiced, followed by a one-hour dual cross-country flight and then a solo cross-country flight covering a minimum distance of fifty miles with two full-stop landings at different airports. The last step to a private pilot license was a successful checkride by a CAA inspector.[11]

Akron Airways, Inc., owned by R. A. "Rudy" Van Devere and operating out of both Akron and Mid-City Airports, was selected to provide flight training for the programs at Kent State University and Akron University. Each college appointed a coordinator, usually a ground school instructor, who worked with the flight instructors. The coordinator was also responsible for handling everything non-flight-training related.[12] At Kent State, this job as KSU CPTP flight coordinator fell to Adrian Van Wyen, a member of the three-person industrial arts faculty.

Applications at Kent State were accepted for the first CPTP class of student pilots, which would include women. President Leebrick promoted the program: "I can't emphasize too much that this aviation course is a real opportunity and those students who want to be considered should register immediately."[13] By the time the enrollment period ended in October, there were eighty-two applications for forty slots. War was breaking out in Europe following Nazi Germany's invasion of Poland the previous month, but the student newspaper, the *Daily Kent Stater,* reassured students that the CPTP "involves no military or other obligation to the government."[14] Although the classroom slots were filled, ground instruction only began in November when instructor Emmanuel Pierret arrived. The seventy-two hours of preliminary ground instruction were structured to give a thorough introduction to the functions of the airplane, its controls, and its instruments.[15]

Flight training was delayed until CAA inspectors arrived to inspect conditions and flight instructors at both the Akron and Mid-City Airports. Also, Van Devere had some hesitation about giving his instructors employment contracts before they passed the rigid CAA qualifications.[16] In November, CAA inspectors approved Akron Airport for instruction for both Kent State University and Akron University students. Mid-City was not approved due to the runway conditions but could be used as an auxiliary airport when too many students were flying at Akron. Five instructors were also approved by the CAA inspectors, but another three were still needed to handle the student load of the two universities.[17]

Finally, on November 30, 1939, the first flight training took place for the KSU students at Akron Airport. Instructor Russell Kirbert got all ten students in his group into the air: Arnold Bereit, Donald W. Dietz, Robert Heighton, Robert Mathews, Robert Schwering, Mary Sherriff, Robert Stueve, Fred Swasey, Wayne Wendell Wilkin, and James McCune.[18]

To prevent future delays in training for KSU students, the board of trustees approved a five-month appointment for Pierret as CAA ground instructor starting in December 1939. That may have fixed the ground school, but the students were unhappy with the location of flight training. With Mid-City

Form 339 (CAA-20-2)
6-1-39

UNITED STATES OF AMERICA
CIVIL AERONAUTICS
AUTHORITY
WASHINGTON, D. C.

Temporary Student Pilot Certificate
(PASSENGER CARRYING PROHIBITED)

This Certifies, *that* Mary Frances Sherriff
of (address) Engleman Hall Kent, Ohio

is rated as a student pilot for the duration of this certificate. The terms of this certificate include each and every provision of the currently effective Civil Air Regulations issued by the Civil Aeronautics Authority pursuant to law. Unless sooner suspended, revoked or taken up by a Civil Aeronautics Authority Inspector, this temporary certificate expires 60 days from date of issuance or upon issuance of the succeeding permanent certificate, at which latter time the temporary certificate shall be affixed to and kept with its successor certificate. Unless otherwise determined by the Civil Aeronautics Authority the holder hereof is now

physically qualified for no higher grade of pilot certificate than that of Commercial CPT Pilot Certificate.

Date issued 10-9-39

By direction of the ~~Secretary of Commerce.~~ Authority

Mary Frances Sherriff
Signature of Student Pilot

H G Ruby
Authorized Medical Examiner
Civil Aeronautics Authority

Any alteration of this certificate is punishable by fine or imprisonment.

Civilian Pilot Training Program student Mary Sherriff's temporary student pilot certificate with address of Engleman Hall, the female-only dorm, October 1939 (Stubbs Family Collection)

Adrian Van Wyen notifies the first group of KSU Civilian Pilot Training Program students that they will receive flying instruction at the Akron Municipal Airport, November 1939. *Left to right:* Robert Matthews, Robert Schwering, Robert Heighton, Donald Dietz, Fred Swasey, Arnold Bereit, Wayne Wilkin, and Mary Sherriff. Absent from the picture are Robert Stueve and James McCune. (Wilkin Family Collection)

available only as an auxiliary airport, students had to go to Akron Airport with no transportation provided and complained that "frequently we have to miss our flying time altogether."[19]

As 1939 drew to a close, there were 9,350 men and women at 435 universities and colleges learning to fly in the CPTP. The program was offered in every state, as well as Hawaii, Alaska, and Puerto Rico. With forty students, Kent State had the largest enrollment in the CAA program in northeast Ohio.[20]

Conditions at Mid-City Airport improved enough that by January 1940, twenty KSU students (Groups 1 and 3) were conducting their flight training

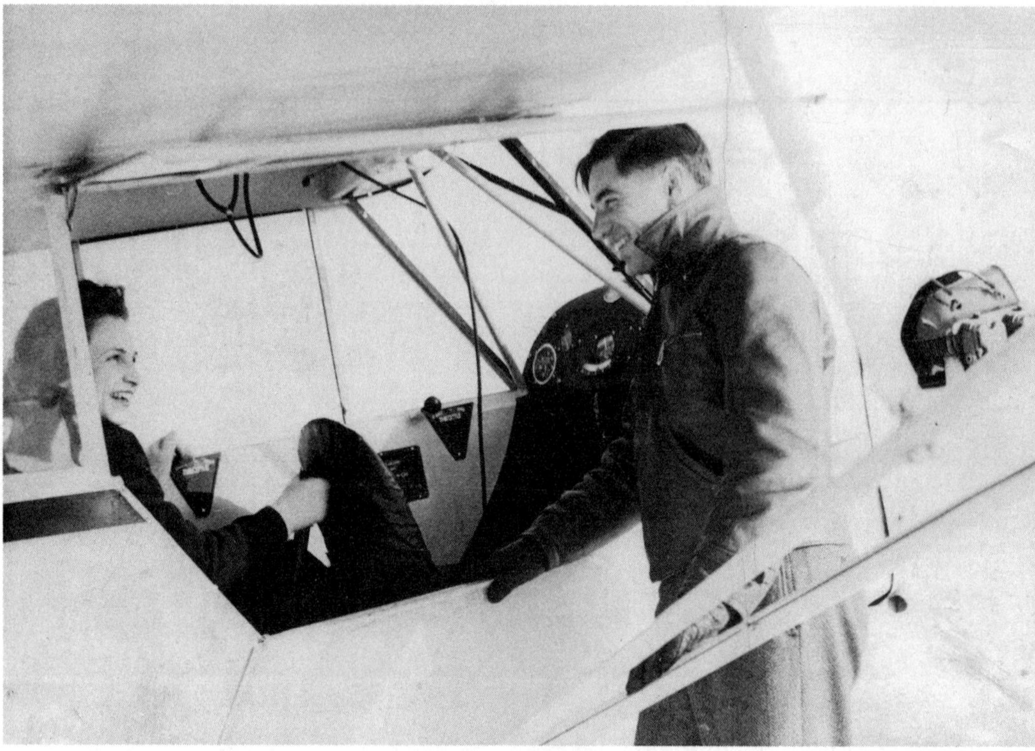

Mary Sherriff, the first female Civilian Pilot Training Program student in the Kent State University and Akron University classes to solo, with her instructor Kenneth Ropp, 1940 (Stubbs Family Collection)

there. The weather must have been good because thirty of the KSU students completed 126 hours of flight instruction in the first week of the new year. The primary trainer was the Piper J-3 Cub with a few flying in a Stinson 105. Ten KSU students had yet to start flight training six weeks after the others had begun because Akron Airways was still short one flight instructor.[21]

On January 11 at Mid-City, Donald Dietz, a sophomore from Liverpool, Ohio, became the first KSU student to solo. Instructed by Kenneth Ropp, Dietz is credited with starting the custom of the student pilot giving a gift to his flight instructor after soloing. Another male student soon soloed, but attention was focused on which female would become the first to follow suit. As January 1940 wore on, the race narrowed down to Mary Sherriff of Kent State flying out of Mid-City Airport and Elynore Fulton of the University of Akron flying out of Akron Airport. Both the *Daily Kent Stater* and the *Akron Beacon Journal* kept tabs on their progress. On January 28, Mary Sherriff soloed at Mid-City. The local newspapers carried news of her accomplishment along with a note that University of Akron had yet to solo any of their female pilots.[22]

By the beginning of February 1940, nearly half of all the KSU students had completed their first solo, including the second woman, Betty Bayes of Olmsted Falls. As February came to a close, the forty KSU students had completed 1,920 hours of ground school and 200 hours of dual and solo flying. Despite harsh winter weather in northeast Ohio and Mid-City's proximity to the snowbelt, instructors and students had managed to fly through all the challenging conditions.[23]

With the arrival of spring, the pace of flight training increased significantly. At this rate, the students would be able to complete their needed hours and earn their private pilot licenses on time (by June 15). Students were moved to Akron Airport for a month while the Mid-City runways were reconditioned for spring training.[24] Deteriorating runway conditions at Mid-City led the students to rename the airport "Mud-City."

Enthusiasm and motivation were high among the KSU students, something noted by Akron Airways' president Rudy Van Devere: "Why even in vacation weeks, the Kent Staters managed to get down here, furnishing their own transportation, while very few Akron U students took advantage of the holidays." More flying was done in one week in the month of April than in any previous week of the program.[25]

By this point, the students in Group 1, who were the first among the four KSU groups to start their flight training in December 1939, were rapidly approaching the end of the course. Everyone in the group had completed the cross-country portion of Stage C and was readying for the final checkride. The first two students who attempted to earn their private pilot license failed, but others would be successful.[26]

First Earned Private Pilot Licenses by KSU Students

On April 10, 1940, KSU students Henry Russell Kenyon and Frank Petee Jr. arrived at Mid-City Airport for their checkrides with three CAA inspectors from Cleveland. Petee climbed into a Piper Cub to start the test. For the first portion, the inspector watched from the ground as Petee performed two spins and a complete spiral landing. The first went well, but Petee admitted he missed the spot where he was supposed to land. With those maneuvers completed, the inspector climbed into the airplane for the remainder. Once in the air, the inspector watched as Petee performed climbs, descents, turns, and the eights on pylon maneuver.[27]

After completion of those maneuvers, the inspector told Petee to fly back to the airport. What probably seemed like a nice gesture at first, Petee was handed a stick of gum by the inspector. While Petee was distracted accepting the treat, the inspector cut the throttle to idle, showing the checkride was not yet over. Now it was up to Petee to demonstrate that he could glide his airplane to a field for a safe landing with the simulated power loss. Accounting for the winds and his altitude, Petee handled the situation as any well-trained pilot would. Finally, it was time to head back to the airport, where he finished the flight with a decent landing. And with that, Frank Petee became the first KSU student in the CPTP to earn a private pilot license. Kenyon was also successful that day and finished his checkride soon after Petee landed.[28]

On April 18, another first occurred. After forty-four hours of dual and solo flight time, Mary Sherriff of Sharon, Pennsylvania, earned her private pilot license. While she was the third KSU student to do so, she was the first woman

Mary Sherriff and Mary Jane Willson with a Great Lakes 2T-1A Sport, 1940 (Stubbs Family Collection)

for the university as well as for the entire state of Ohio. Her achievement was widely reported in both local newspapers and those from bordering states.[29]

Despite Sherriff's accomplishment, women in the CPTP did not have it easy. Akron Airport manager B. E. "Shorty" Fulton stated that only one woman in a thousand could be as good a pilot as a man, and some of the instructors fell in line with that thinking. Various reasons were suggested. One Akron instructor reasoned it was because "you can't bawl out a girl student like you do a boy." Others believed it was because women were not mechanically minded or that they lacked nerve. Fulton was quoted as saying: "Why I wouldn't think of flying a plane if I didn't like to tinker around mechanical things. You have to go a long ways these days to find a girl with the record of a couple of hundred flying hours to her credit. Now if our girls were Amazons like the Russian women . . ." Interestingly, his own daughter, Elynore Fulton, was a CPTP student at Akron University.[30]

Van Devere believed that all the student pilots were held to standards as tough as those for transport pilots, but the female students "just don't seem to take the bull by the horns quite so quickly. They don't gain confidence to fly by themselves." Meanwhile, Art Stanley, Mid-City manager and instructor, saw it as a lack of nerves: "I think it's because women are not as daring as men." Mary Sherriff was considered by some to be an exception.[31]

Dealing with biased instructors did not stop women from taking advantage of the CPTP. In the previous two and a half years, the number of women holding licenses jumped from 444 to 761. Of those, 429 held a private pilot license, while nearly 100 held commercial or transport licenses.[32]

Aviation Expands at Kent State University

On March 26, 1939, a group of thirty-one young men, mostly employed by Black and Decker in Kent, Ohio, formed Kent Flying Club, which operated out of Stow Field. Their club became the most active private pilot organization in the area. More than half of their members were married and nearly as many were under the age of twenty-five. For most, this was their first introduction to flying. Nearly all had soloed by April 1940. Robert Dix, publisher of the *Kent Courier-Tribune,* was the only member with an airplane. The rest rented a Piper J-3 Cub. The club had a rule that each member had to fly at least half an hour a month. The flight instructor for the club was Art Laurence, who was also the manager at Stow Field.[33]

In its first year, the club reported that members logged 600 hours at Stow Field without an accident while flying after work or on weekends.[34] In the summer of 1940, Laurence left for a new job at the Ryan School of Aeronautics in California. At the time, Laurence had over 5,000 hours of flight time and had soloed most of the club members.[35] Acknowledging female interest in aviation, the club announced the formation of a women's auxiliary to which members urged their wives, girlfriends, and any other interested women to join. It took over a year to get the auxiliary up and running, and the club finally announced women were eligible to join in November 1941.[36]

CPTP flight training flourished with the return of better flying weather. After a month of rolling and grading the runways at Mid-City, Van Devere reported its condition as being in "better shape than it's ever been." Early in May 1940, KSU students would be able to return to Mid-City for their flight training. Each week brought more students earning their private pilot licenses, drawing their involvement with the CPTP to a close. To mark their successes, the students and instructors held a banquet on May 9, 1940, at the Hotel Kent to conduct "farewell graduation exercises." By the time the first group of forty students completed the CPTP, thirty-eight had earned their private pilot licenses. This class of forty was the third largest in Ohio, behind only Ohio State University and University of Cincinnati.[37]

Given this student success, Van Wyen proposed that Kent State University reapply for the program for fall quarter along with requesting the new advanced course that had 146 hours of ground school and 50 hours of flight instruction. The course focused on advanced work on engines and aircraft, navigation, radio use, night flying, cross-country practice, and maneuvers. A certain percentage of those who completed the primary course would be eligible for the advanced training. Van Wyen found that KSU students "are considerably farther advanced than many other schools." The board of trustees considered the proposal at their May meeting and authorized President Leebrick to apply for continuation of the training program and the advanced training program.[38]

The increased flight training began causing traffic congestion issues at Akron Airport. In addition to the seventy CPTP students, there was more private flying due to the improved weather and additional airline flights. Local model builders had taken to using an area behind the Goodyear Airdock, which resulted in a few instances where model flying interfered with landing aircraft. Fulton erected a sign advising the junior aviators to get a schedule of aircraft activity at the airport before flying. Because of these traffic issues, officials were stationed in the new control tower to manage traffic."[39]

By early June 1940, Kent State received approval from the CAA to offer both the primary course and new advanced course, also called the secondary course, during the summer. Due to the success of the CPTP nationwide,

Civilian Pilot Training Program students flew the Waco UPF-7 at the Akron Municipal Airport. A ground crew member starts the engine for a student and instructor by hand propping, 1940. (Kent State University Libraries. Special Collections and Archives)

the government was eager to get as many students enrolled as possible. The student quota was increased to sixty. The number of instructors at both Akron and Mid-City was to remain the same, which meant each airport was going to be responsible for fifteen students rather than ten. More airplanes were needed to hold to the requirement of one airplane per ten students. Women were still given 10 percent of the available slots.[40]

The national goal set by the CAA was for 45,000 primary and 9,000 secondary students to start in July. To help train these pilots, 3,000 additional flight instructors would be qualified, as well as 5,000 refresher courses given to pilots with extensive experience and lapsed licenses. With the addition of the advanced training course, Van Devere received notification to send one instructor to Chicago for two weeks of advanced training under CAA rules. Once that training was completed, the instructor would be able to give refresher training to pilots who had let their licenses and knowledge of CAA rules lapse.[41]

To increase the pool of eligible students, the program was expanded to include high school graduates who had passed their college entrance exams and any student who had completed two years of college training or one year of college in 1939–40. Priority was given to students currently enrolled at Kent State, then to students from other colleges, and last to high school graduates. In addition, the financial burden was lifted for students. The physical exam was free, although failing to pass would cost the applicant six dollars. The previous course fee was dropped, making both the ground school and flight training free. The federal government would pay room and board and also arrange transportation if necessary for students accepted from a town with no training program.[42]

Akron Airways was one of one hundred schools in the country selected to offer the advanced training. It had fifteen airplanes in its fleet and ordered six more. With the expected increase in CPTP students, the CAA chief inspector for private flying made a trip from Chicago to inspect possible sites at Akron Airport for student barracks.[43]

To help meet the CAA goals, Akron Airways started a course to train new flight instructors that consisted of ten to fifteen hours of flight and ground instruction financed by the government. To be eligible, a pilot had to hold a commercial license and be twenty-one to forty years old. The biggest competition for qualified pilots came from the airlines, with their lure of better pay. To compete, the government increased the starting pay of flight instructors to $175 to $200 a month. Eventually, their pay would be as high as $300 a month, comparing favorably with airline pay.[44]

During the summer of 1940, United Airlines visited schools operating the CAA program to interview students in the secondary training course interested in flying for the company. Upon selection, the pilots were given a scholarship to attend the Boeing School of Aeronautics in Oakland, California, for training to become a copilot with United Airlines. The company made a trip to Kent State and selected two high-ranking students, Wayne Wilkin and Robert Schwering, who joined the inaugural class of forty pilots.[45]

The continued success of flight training drew the attention of the US Navy. At the same time CPTP enrollment increased, the US Navy submitted a request for ten students to be recruited from both Akron and Kent State Universities for an assignment into the navy's lighter-than-air classes. Van Wyen hoped for at least forty students to apply for this training. A prospective student had to be twenty to twenty-six years old and have college credits in algebra, plane and solid geometry, trigonometry, and physics. Training would be held at Grosse Isle, Michigan, and Lakehurst, New Jersey, where cadets would be paid $84 and $105 per month, respectively.[46]

By the end of July 1940, the first class of CAA advanced training began at both Akron and Mid-City Airports. Akron Airways' instructor Sam Williams had finished his training in Chicago and now had twenty students from Kent State University and Akron University who would receive their training at both airports.[47]

Coinciding with the launch of the advanced course, Van Devere took an open-cockpit Stearman Model 6H Cloudboy for a test flight, only to find himself thrown from it during a maneuver. As Van Devere rolled the airplane inverted at 2,800 feet in the air, his safety belt broke, and he fell out. "It was a couple of seconds before I remembered my parachute and pulled the ripcord. I fell into a grove of trees, and my parachute hung me about eight feet from

The Stearman Model 6H Cloudboy at the Akron Municipal Airport from which Rudy Van Devere was thrown free when doing a half-roll and half-loop maneuver and survived, 1940 (Van Devere Family Collection)

the ground. I cut the shrouds and dropped." The airplane landed twenty feet short of a house in Ellet, then bounced into it, breaking a few windows. Luckily, neither Van Devere nor anyone on the ground was injured.[48]

An Air Safety Board inspector found the safety belt webbing on the former US Army airplane broke due to rot around a metal fastener. Van Devere was cleared of any blame for the accident and would become known for this incident as one of the few people to fall out of an airplane and live to tell the story.[49]

The fall quarter of 1940 saw the third offering of the CPTP to KSU students, and this time there was an increased focus on military flying. CAA head Robert Hinckley discussed various plans with the army air corps and navy to stimulate voluntary enlistment and make the CPTP more responsive to national defense. In September, all students accepted into the program had to pledge themselves to enter military aviation if qualified and needed for national defense. The army air corps shipped two military Pratt and Whitney engines as well as various aircraft and engine instruments to Kent State to help build interest in military aviation.[50]

Hinckley reflected in early 1941 that until July 1940, the CPTP "was built upon a civilian philosophy, as part of the development of civil aviation." Noting that the philosophy had recently changed, he advised that "our trainees should be placed in some definite status related to the armed forces before we begin their training."[51]

At the beginning of the fall quarter, the army air corps sent an air officer to visit Kent State, Akron University, and Akron Airport to recruit men. Any primary or secondary student or any man with two years of college education who could pass the physical examination was eligible. Rules, however, were going to change for men interested in joining the army air corps. Selective service registration was about to begin on October 16. Van Devere received a letter that indicated that thereafter many men wanting to get into the flying service would not be eligible under the new registration system.[52]

The CPTP quota for fall quarter 1940 was forty students, yet at the start of training there were still fifteen vacancies despite articles in the *Daily Kent Stater* reporting available slots. This was the first time that CPTP enrollment was down at Kent State. It is not clear if this was because of the requirement to sign a pledge for military enlistment, if the pool of available and able-bodied students was drying up, or if there was just a lack of interest. By the time the enrollment window closed, a few more students were added to the ranks, and three groups of ten began flying at Mid-City Airport.[53]

Efforts had been made at Kent State over the previous year to establish a Reserve Officers' Training Corps (ROTC) program with no success. Ohio congressman Dow Harter felt that in time a ROTC unit would be created at Kent State, but that now there was a lack of army personnel to do so. He also foresaw the CAA program continuing for years with little guarantee that the students involved would ever end up flying for the army air corps. The lower enrollment in the CPTP at Kent State was not unique among schools across the United States. Congressman Harter's words were probably an attempt to reassure potential student applicants that it would not be a direct path to fighting in a foreign war.[54]

For the last few years, Stow Field had seen little use. The rise of CPTP flight-training activity at nearby Akron and Mid-City Airports had negligible impact on flying at the Stow airport. The airfield had been in foreclosure since May 1936 with a court order for a sheriff's sale.[55] There was some flight training provided for the public starting in 1937 that continued in a limited way for several years. Most flight activity had been limited to that of Kent Flying Club.

Rudy Van Devere Purchases Stow Field

A big change came to Stow Field in late September 1940 when Van Devere bought 81.55 acres of the property, including the house. He had plans to run the airport primarily for private fliers but saw the possibility of bringing in some of the CPTP student operations. Van Devere did not plan to improve the field conditions until the following spring. Given the conditions at Stow Field, this delay ensured little to no flying would occur there until sometime in 1941.[56]

The CPTP at colleges and universities was a boon to aviation in Ohio. War preparedness had helped the economy show its first sign of recovery from the Depression. Van Devere suggested in December that Kent State buy Stow Field from him, prompting KSU President Leebrick to begin negotiations to bring Stow Field under university ownership or at least its direction. This was the first indication that Kent State saw aviation as having a more permanent place at the university.[57] Although the university's first attempt to buy Stow Field, which would require funding from the state, was not successful, it did not diminish the university's desire to have its own airport.[58]

Growth in the aviation industry was also putting strains on available manpower resources, and the lack of qualified instructors was a grave problem. Pilots with the required hours were leaving for Canada, where the government offered more money and training. For that reason, the CAA started offering a commercial refresher course. The course was open to any pilot with enough

Akron Airways' Civilian Pilot Training Program instructors with Piper J-3 Cubs used for student training at the Akron Municipal Airport, 1940 (Van Devere Family Collection)

hours for the commercial license but who had never taken the test to earn it. Upon finishing the course, a pilot would be qualified as an instructor.[59]

Akron Airways started offering a new apprentice flight instructor course. While not directly connected with any college or university, a prerequisite was successful completion of the restricted commercial course being offered at Kent State for the first time. A restricted commercial license meant the pilot had completed the required training for the commercial license but did not meet the total flight time requirement. As soon as the pilot logged the required two hundred hours of flight time, the license would no longer be "restricted." The apprentice flight instructor course included twenty-five to thirty-five hours of flight instruction and twenty hours of ground school. Upon completion, the apprentice instructor was given five CPTP students to train. Once all five students passed their private pilot checkride, the apprentice was given his flight instructor rating.[60]

Interest in Aviation Begins to Wane

The growth in aviation in the Akron area in 1940 was impressive. CPTP students at Akron and Mid-City Airports increased from 280 to 736. New flight courses for advanced students and new instructors were added at both airports. The number of instructors teaching increased from seven to twelve, and the number of mechanics tripled. Other small airports in the area were experiencing increased business, but that was not the case at Stow Field. Little had improved there since Rudy Van Devere bought the airport in September. The sale to Kent State had fallen through, but improvements were expected to make the airfield usable for private pilots in the spring.[61]

The CPTP was offered for the fourth time at the beginning of 1941. The initial quota of forty primary students and ten secondary students was dropped to thirty when only eleven students had applied after a week. Seeing the University of Akron's program get cut the month before may have seemed like a warning to Adrian Van Wyen. After conferring with Leebrick, Van Wyen announced that women were permitted to enroll in the primary course.[62]

Van Wyen received notice in January 1941 that the US Navy was looking for pilots. While the navy was busy looking for college applicants, so was Kent State. Nearly a month into the quarter, the quota for student pilots was still not filled. Within Ohio, Kent State was in the top third of more than a dozen colleges and universities in the training of student pilots. Even then, interest had severely dried up at the school.[63]

With the war in Europe continuing to escalate, there was still strong opinion in the United States favoring isolation from world conflicts. That, along with the CPTP's shift to more military preparation and waning student participation, led to questioning on US campuses of the need for the program. The future of the CPTP would be decided by federal funding. In February 1941, the budget was cut from $37 million to $18 million. The CAA plan was either to cut quotas at schools or drop the courses completely. Given that Kent State was clearly having trouble filling its quota, Van Wyen saw that the future of the CPTP at the university was in jeopardy, which led the CAA to change policy and allow freshmen to apply to the CPTP.[64]

By the time the primary course began at Mid-City in the beginning of March 1941, enrollment had increased from eleven to twenty, still far from the original quota of thirty students. Van Wyen was notified by Washington that Kent State's CPTP was canceled for the summer, but that Kent State should expect the courses to resume in the fall. On the Kent State campus, Leebrick held a special assembly to address the university's support of a national defense program. In addition to participating in the CPTP, "the University has already taken various steps in the preparation with the CAA and with attempts to get a unit of the naval ROTC." Industrial arts professor Clinton S. Van Deusen stated, "We must change the phrase 'keep America out of war' to 'keep war out of America.'"[65]

In the spring of 1941, Stow Field still saw little to no traffic. Perhaps as a sign of how quiet it was, Van Devere moved his family with two young children in May 1941 from an apartment above the Akron Airways' hangar at Akron Airport to the big house at Stow Field. He turned the apartment they left behind into a club room for Akron Airways' student pilots. Van Devere's son Joe was about ten years old when the family moved to Stow Field. He believes his dad might have been motivated to buy the property for the house so that the family could move out of its quarters above the hangar at Akron Airport, where they had been for four years. He remembers that the existing hangar at Stow Field was in shambles, "a piece of junk." In 1941, Van Devere had been given approval to establish a training school for air cadets in Helena, Arkansas, and was regularly commuting to Stow on weekends. By January 1942, he had fixed up the old hangar at the airport to use as a "garage" for his own airplane.[66]

State Funding for Five University Airports

At that same time, a meeting took place in Columbus, Ohio, that would ensure a future for collegiate aviation in the state. The heads of five state universities were called to Columbus, where a proposal was made that the State of Ohio would acquire airports in their name.[67] The universities in attendance were Ohio State University, Ohio University, Miami University, Bowling Green University, and Kent State University. The airports were to be located close enough to the universities to provide for the instructional needs of both the present and the future.[68]

Ohio State University had been using the Columbus Airport for its CPTP and had limited options for buying a small local airport. Its solution was to buy a field and start from scratch. Kent State had been using Mid-City for training for the last year and a half, but it was not the closest option to the school. President Leebrick was reportedly looking at Stow Field to become the official airport for the university. Stow Field seemed a perfect fit, and Kent State had shown interest in buying it once before. Nevertheless, the university considered other options, and it would take over a year before Kent State bought Stow Field.[69]

On the heels of this development out of Columbus came another big change out of Washington. It was announced that starting in the summer of 1941, women would no longer be accepted into the CPTP. This decision drew the ire of newspapers, women's organizations, and even Eleanor Roosevelt. The CAA issued a statement explaining that they had been given a directive from congressional officials to change their program to exclude females as "male pilots have a wider and more varied potential usefulness to the armed forces than female pilots." Since its inception, the program had increased the number of women pilots in the United States from 675 in 1939 to nearly 3,000 by July 1941. The exclusion of women did not just fall to the CPTP. Ray Van Devere, Rudy's brother, announced a "$80-to-solo" course at Akron Airways that was "definitely not open to women students."[70]

The CPTP's focus now moved completely to military preparation.[71] The push by the CAA for more pilots meant Kent State rescinded its cancellation of the summer program. At the end of May, Van Wyen announced summer primary and secondary courses with a quota of ten students for each, but by the time summer arrived, neither course had met its quota, even though they were open to students from other colleges or universities.[72]

Another blow to local CPTP training came in September 1941 when Akron Airport announced it would no longer serve as a training facility due to military aircraft traffic congestion. This followed the removal of private flying six months earlier, again due to traffic. Akron Airport's manager Shorty Fulton also started requiring all student pilots to use radios, which Van Devere cited as one of the reasons for the move: "The students aren't used to radios and frequently can't get messages."[73]

KSU students were already doing most of their flight training at Mid-City, but now they were joined by the University of Akron students. However, CPTP training would still take place at Akron Airport for those students who required night flying, as that was the only lighted airport in the area.[74]

The start of the fall quarter at Kent State saw an increase in student interest in the CPTP. With twenty slots available, eighteen students signed up for the primary course. Perhaps a visit from CPTP alumnus Robert Schwering helped inspire students. Schwering was one of the first forty students to start in the program in 1939 and was now a copilot with United Airlines. He shared stories of his experience and of other KSU CPTP graduates.[75]

In fall 1941, the curriculum committee of Kent State's liberal arts college authorized credit for two CAA ground school courses. This marked the first time that college credit was given for an aviation class by Kent State. Aeronautics 260 and 360 were open to all students, including women, who were sophomores or juniors. There was no requirement to pass a physical or be enrolled in the CPTP.[76]

Andrew Paton Hired, Stow Field Remains Idle

As the fall quarter ended, Van Wyen recommended Andrew W. Paton to Leebrick for a new part-time instructor position. Paton had been at Kent State in 1941 working on his master's degree. While not yet a pilot himself, he had completed the ground instructor course at Kent State during the summer. He was hired and began teaching on January 8, 1942.[77] Over the next two decades he would be instrumental in the development of Kent State's aviation curriculum and serve as an advocate for the airport as an educational facility.

No decision had been made regarding the acquisition of an airport by Kent State nearly a half a year after the state's offer to buy one for the university. With no airplane traffic, Stow Field became completely overgrown. Van Devere had yet to make anything of his airport purchase but was still hopeful that he could sell it to an interested party once he fixed it up and made improvements. One solution was to turn the airport over to private fliers at cost. They would operate and care for the grounds while he would sell them the needed fuel and oil. However, with other well-maintained airports nearby, it was going to be a tough sell.[78]

Andrew W. Paton (William Andrew Paton Family Collection)

Andrew W. Paton (1908–1964), a North Dakota native, moved to Chardon, Ohio, as a young boy. He married in 1930 and graduated from Wooster College the next year. For ten years, he taught industrial arts and science in Geauga County. Paton then moved to Kent for his job at Goodyear Aircraft Corporation, writing manuals for the Corsair fighter, K-class blimps, and B-26 engines. Kent State hired him in 1942 as a part-time CPTP instructor and in 1945 as an industrial arts faculty member, charging him to develop its aviation technology curriculum. Working with interested students, Paton helped form the Kent State University Flying Club and served many years as its advisor. His flight certificates included ground school instructor and commercial pilot, and his logbooks suggest that many Columbus flights were to the state surplus warehouse. As early as 1951, Paton advocated that Kent State take over flight training and expand the airport's educational facilities. His 1952 master's thesis recorded the early years of KSU aeronautics training. In 1959, Paton took a one-year leave to train industrial arts teachers in the Philippines. He was a full professor and served as aviation technology coordinator and airport manager until his untimely death in 1964. He was recalled as a man whose "influence for good is immeasurable." His legacy was his vision for establishing flight training at Kent State and promoting the airport as an educational facility. The airfield at the Kent State University Airport is named in his honor.[79]

Summary

As 1941 drew to a close, the United States was not yet a combatant in World War II, yet the nation was clearly engaged in the war effort. The CPTP had fulfilled one of its original goals to produce a pool of pilots that the US military could call upon. The aviation industry had been rejuvenated as a result, with the manufacture of training aircraft and airport improvements to accommodate the traffic. While the desire to remain isolated from foreign conflict still existed, it would not be long before war came to the United States' doorstep.

It may have been personal convenience that led Rudy Van Devere to purchase Stow Field, but that action preserved it through the prewar years. In hindsight, had this purchase not been made, Stow Field most likely would have disappeared.

Chapter 6

Wartime at Kent State University, 1941–44

By 1941, the Civilian Pilot Training Program (CPTP) was present in the forty-eight states and US territories, including Hawaii. Many of its trained pilots would find themselves drawn into the war. Two CPTP graduates from Oahu would be in the first airplanes shot down in the Japanese attack on Pearl Harbor. Sgt. Henry Blackwell and Cpl. Clyde Brown chose to take one last flight in Hawaii on the morning of December 7. Along with a friend, they took off from John Rogers Airport at Kapolei, Hawaii, in two rented yellow Piper J-3 Cubs at 7:40 A.M. Initially the two pilots flew east past Pearl Harbor, then south along Waikiki Beach toward Diamond Head, before turning their airplanes westward. A sizeable group of aircraft approached quickly from the west. A navy shipman on a US Navy tugboat leaving the channel at Pearl Harbor saw the two small Cubs shot down by the approaching unidentified aircraft. Neither the Cubs nor their occupants were ever recovered. The two CPTP graduates and their passenger were the first American military personnel killed by the Japanese in World War II.[1]

Pearl Harbor Attack

Aviation in the United States changed immediately following the attack on Pearl Harbor. That same day, the Civil Aeronautics Authority (CAA) issued Order Number 3, which suspended all pilot certificates as of December 8, 1941, until later consideration for reinstatement. It also requested that police officers be assigned to all landing fields to protect the facilities and that aircraft be grounded unless scheduled for air transport or government operation. That same day, the CAA authorized its inspectors to seize or impound any aircraft owned, operated, or piloted by an alien, a person without a pilot license, or a holder of a pilot license that had been suspended by Order Number 3. Airline passenger baggage was required to be inspected by pilots before departure starting on December 10, and cameras were no longer permitted in the aircraft cabin.[2]

In Summit County, Ohio, alone, these new regulations affected about 200 licensed pilots, in addition to the students participating in the CPTP. KSU CPTP coordinator Adrian Van Wyen received a letter from the CAA on December 9 that all CPTP flying would cease until further notice. In addition, all students were ordered to produce birth certificates and written recommendations from three responsible persons. Those recommendations were then to be given to the CAA commissioner to determine if the student had unquestionable loyalty to the US government. Fingerprints and a photograph were also new requirements. CPTP students were informed that flight training and academic coursework were now their sole, full-time job.[3]

Less than a week after the attack and the immediate grounding of all nonmilitary and air transport flying, the CAA limited the number of people who could receive a reissued license. While 80,000 private pilots were grounded in the United States, steps were quickly taken to resume the CPTP. The CAA issued an order stating: "Until further notice, only CPTP instructors, CPTP students and pilots working in the interest of national defense may apply. Pilots not in this category are requested not to apply at this time or call CAA offices." Flight instructors were given priority to obtain reissued licenses, followed by students. Five days after the attack, Akron Airways, Inc., had three flight instructors reinstated, along with five students, followed by another fifteen students a few days later.[4]

Even prior to December 7, the CPTP had increased its military emphasis. Once the United States declared war, the future of the CPTP completely shifted to training military pilots. President Franklin D. Roosevelt issued Executive Order 8974 on December 12, transforming the CPTP into a wartime program. All civilian pilot training would be "exclusively devoted to procurement and training of men for ultimate service as military pilots or for correlated non-military activities." It soon followed that the CPTP was renamed the War Training Service, but the name change did not become official until a year later.[5]

Airports were required to have at least two guards providing twenty-four-hour security to prevent spies or saboteurs from getting access to any aircraft. If an airport could not meet this requirement, it was closed. Stow Field

Form ACA 935
DEPARTMENT OF COMMERCE
CIVIL AERONAUTICS ADMINISTRATION
WASHINGTON

This Is To Identify
Wayne Wendell Wilkins
(First name) (Middle name) (Last name)
who has been certificated as an airman by the Civil Aeronautics Administration and whose signature appears hereon.
Signature

Signature James P. Colton
Civil Aeronautics Representative.
Signature James P. Colton
Fingerprinting Officer.

This certificate is valid until recalled

Place of birth Akron, Ohio
Date of birth November 29, 1917
Color eyes Blue hair Brown
Weight 155 lbs. Ht. 5 ft. 9 in.
FINGERPRINTS—RIGHT HAND
RIGHT THUMB

License of Civilian Pilot Training Program student Wayne Wilkin following the attack on Pearl Harbor, December 1941 (Wilkin Family Collection)

somehow survived this period of airport shutdowns. While it was a little-used airport that had played no role in CPTP training, it had the strong Rudy Van Devere connection that may have made the difference. In any case, while other airports were lost, Stow Field was kept open during this critical period.[6]

An order from Washington said aircraft kept at unguarded airports "must be rendered un-airworthy immediately by the owner or his agents by removal of some parts of the essential mechanism." The CAA also required that all aircraft be registered, recording the owner, type of aircraft, its size, and the airport or place of storage. The purpose was to form a database that could be used in a time of emergency. The government took another measure to thwart illicit aviation activities by issuing a temporary wartime ban on the release of weather data from US weather recording stations. A limited one-day forecast was given daily with a twelve-hour forecast at 7:30 A.M., followed by a nineteen-hour forecast at 11 P.M.[7]

By the start of 1942, private pilots outside of the CPTP had started to get their reissued licenses. They still were not allowed to fly, and the rules remained in a constant state of flux. Upon reinstatement, pilots were made part of the new Civil Air Patrol.[8]

On February 15, 1942, all private pilots were allowed to resume flying, albeit under a considerable number of new regulations and procedures. Charles I. Stanton, acting CAA administrator, said: "Our prime objective was to 'Keep 'em Flying.' We work on the theory that every pilot and every plane in the nation is an asset in the total war effort, and that we should make rules on the basis of what a pilot can properly do rather than what he cannot do."[9]

War Years for the Civilian Pilot Training Program

Immediately following the Pearl Harbor attack, students who had completed their CPTP course began enlisting in the armed forces. While enrollment had been suffering the year before, now there was increased interest among KSU students to learn to fly. Colleges were getting more applicants than their quotas could accommodate. CAA radio advertisements for training spurred interest but were not well received by all. Dean Ayers at the Akron University wrote a letter criticizing the CAA for their advertising and the strain it put on Akron's limited class quota. Kent State's quota for the spring quarter doubled from ten to twenty students for the primary course.[10]

As with the rest of aviation, there were plenty of new rules for those looking to enroll in CAA pilot training. New policy required applicants to apply for the CPTP initially and then, upon being accepted, enlist in the army or navy reserve. They were then placed on inactive status for the remainder of their flight training. The required age limit for applying dropped from nineteen to eighteen years. Married men were eligible if their dependents had other means of support. They were required to take the advanced flight course and also had to sign an affidavit that they would apply to the armed air services or at least the armed forces.[11]

CPTP students were in a unique position at Kent State University once accepted. They were not matriculated as students and therefore not subject to university rules and regulations, nor were they responsible to the university authorities. With their inactive status as a reservist in either the army or navy, they were not officially under the discipline or restrictions of the US armed forces. Kent State found itself providing room and board for college-age enlisted men who had no formal military oversight. It was a situation that caused its share of problems. Most issues resulted from overstaying weekend leaves, occasional insubordination, and army versus navy rivalry. Ultimately, students could have been dismissed from the program and sent to a different branch of the military, but that never happened. Troublemakers were punished with marching, airplane washing, cleaning duty, or loss of weekend privileges.[12]

Uniforms were encouraged by the CAA following the outbreak of the war. At first, students were given a small fabric insignia, which they could attach to their clothing. By July, Kent State had purchased khaki uniforms that they resold to the students at a sizeable loss. Finally, in December, the War Department made arrangements through the CAA to ship forest green winter outfits to the university for the students.[13]

Differences in policies between the army and navy regarding their enlistees were a source of problems as well. While those enlisted in the navy air corps received regular enlisted pay, those in the army air corps received nothing. Navy enlistees were guaranteed cadet training, leading to a commission. Army enlistees were denied regular air corps training or any combat pilot training. Instead, their future service was mostly limited to noncommissioned positions as glider pilots, ferry pilots, liaison pilots, and flight instructors. The navy would periodically visit campus to supervise their trainees, as well as hold conferences with them and the officers. In addition, they required regular progress reports from the university about their trainees' progress. The army provided none of this oversight for their trainees.[14]

Since the inception of the CPTP, 15,000 graduates nationwide had joined the army and navy air services, with another 9,000 going into other branches of the armed forces. In the months following Pearl Harbor, one-third of new army and navy flying cadets came from the CAA program. Of the cadets that came from the CPTP, 11.8 percent failed or washed out of their army flight training. Conversely, 43.4 percent of the cadets who had not participated in the CAA program were unable to pass the training. It was clear that the army saw value in the program and what it provided to the young men joining their ranks.[15]

Recognized for his work as Kent State's CPTP coordinator, Adrian Van Wyen was invited to Washington, DC, in March 1942 to be part of a ten-member committee headed by John P. Morris, the national director of civil pilot training. From this meeting came a plan to expand the CAA pilot training program. Previously, the capacity for training had been 25,000 students per year. The new plan raised that number to 45,000 students. To accomplish this, all CAA pilot training facilities would be devoted completely to the war program in coordination with the army air forces. Students who could meet the requirements as aviation cadets for the army air corps would be given priority. Those who could not meet the requirements were to be trained as flight instructors. Every incoming student had to agree in writing that they would serve the national interests in some fashion in the field of aeronautics.[16]

Initial Training for Army and Navy Cadets

Participation in flight training at Kent State was steady during the last half of 1942. The students were enrolled as army or navy reserve pilots and would either be assigned to complete their advanced training at Kent State or

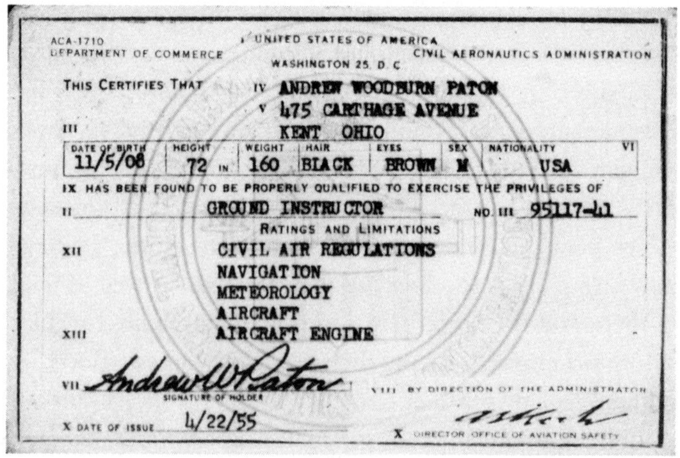
ACA-1710
DEPARTMENT OF COMMERCE
UNITED STATES OF AMERICA
WASHINGTON 25, D. C.
CIVIL AERONAUTICS ADMINISTRATION

THIS CERTIFIES THAT IV ANDREW WOODBURN PATON
V 475 CARTHAGE AVENUE
III KENT OHIO

DATE OF BIRTH	HEIGHT	WEIGHT	HAIR	EYES	SEX	NATIONALITY VI
11/5/08	72 IN	160	BLACK	BROWN	M	USA

IX HAS BEEN FOUND TO BE PROPERLY QUALIFIED TO EXERCISE THE PRIVILEGES OF
II GROUND INSTRUCTOR NO. III 95117-41

RATINGS AND LIMITATIONS
XII CIVIL AIR REGULATIONS
NAVIGATION
METEOROLOGY
AIRCRAFT
XIII AIRCRAFT ENGINE

VII Andrew W. Paton
SIGNATURE OF HOLDER
VIII BY DIRECTION OF THE ADMINISTRATOR
X DATE OF ISSUE 4/22/55
X DIRECTOR OFFICE OF AVIATION SAFETY

Andrew W. Paton's license issued by the CAA showing ground instructor ratings, 1955 (William Andrew Paton Family Collection)

shipped to other training facilities. Starting with the summer session, the total number of ground school hours rose from 108 to 250 hours, and the program was condensed into an eight-week period. The increased program demands led Van Wyen to recommend that Andrew W. Paton be given a full-time position teaching ground school for the next year. Paton had already been teaching part-time, in addition to working at Goodyear Aircraft Corporation. He accepted the offer. During the CPTP and the War Training Service program, Paton would teach ground school for all primary and secondary subjects, plus radio code. To prepare for their army and navy basic training, the students met twice a day on Rockwell Field for calisthenics, military drills, track activities, and wall scaling.[17]

Van Wyen would again be recognized by the CAA for his contributions and appointed as a "Dollar-a-Year Man." US law forbade the government from using unpaid volunteers, so a small nominal salary of one dollar was offered. Van Wyen was to serve as a government advisor for CAA pilot training activities in the region, becoming the first KSU faculty member added to President Roosevelt's Dollar-a-Year Man list.[18]

During the summer, the students were initially housed in Moulton Hall on campus, but later the university moved them to a dormitory owned by Davey Tree Company on North River Street in Kent. Student complaints led to the students being moved to Lowry Hall on campus, which then served as the dormitory for aviation trainees throughout the remainder of the program.[19]

By July 1942, CPTP students were finishing their training and were eligible to start their military careers. The navy had a system for handling its graduates. Prior to graduation, a navy enlistee received his orders, his assignment to a new training center, as well as travel arrangements. These assignments were received with great pride and enthusiasm. The army, however, had no advanced plans for any of their graduates. It was not uncommon for an army enlistee to wait months to hear anything regarding his future training. One particular graduate spent five months as a ground instructor at Kent State, and only after intervention by his congressman did he advance to transport pilot training. This apparent indifference by the army increased dissatisfaction among the army enlistees and caused multiple supervisory problems.[20]

Almost a year to the day after the attack on Pearl Harbor, the CAA stopped recruiting from colleges, and the enlisted reserve became a closed pool. No longer would students in the CPTP come from the ranks of university students.[21] The CAA launched a nationwide campaign for recruits, which garnered a huge response, thanks in part to news of the US invasion of North Africa, filling the set quotas twice over. The CPTP officially became the CAA War Training Service (WTS) on January 15, 1943. The new name reflected that the program was entirely devoted to the training of army and navy pilots, presenting a truer picture of mission.[22]

The start of 1943 brought an increase in student pilots to Kent State. Although the CAA was no longer recruiting from campuses to fill the classes, the incoming army and navy students were still college students. Kent State saw the arrival of forty WTS trainees from schools such as Ohio State University, University of Michigan, Youngstown State, Akron University, University of Chicago, Toledo University, Miami University (of Ohio), and the Naval Academy. The incoming students saw Kent State as a "great outfit" and "very hospitable," while the "social life has far too many limitations for our short stay here." Efforts to ready what was now the Kent State University Airport were hampered by winter in northeast Ohio, so flight training would remain at Mid-City Airport until conditions improved.[23]

End of the CPTP and WTS

As the new trainees arrived at Kent State, the last twenty CPTP students graduated on March 1, 1943. Kent State was notified shortly afterward that the WTS program would be discontinued at the end of that month. At that time, army trainees still in the program were required to report to the local draft board for reclassification. Navy trainees were transferred to other colleges to complete the remainder of their training.[24]

A major difference between CPTP and WTS was that graduates of the latter did not earn a CAA license. That changed in March 1944, when the CAA announced WTS graduates would be granted a CAA private pilot license without any additional flight time or written exams. Graduates who also completed the WTS cross-country and Link Trainer instrument courses and had 160 hours of flight time were granted commercial pilot certificates. This granted new CAA licenses to over 4,100 young men who had graduated from the WTS program the prior year.[25]

Since Kent State started the CAA pilot training program in October 1939, 345 students had graduated with their wings with no serious injuries suffered. Of the graduates, 288 became members of the army and navy flying services. For that period, the total number of pilots on record in the United States grew from 21,000 to more than 100,000 private pilots. While the national average of "wash-outs" in flight training was 5 percent, not a single pilot washed out of Akron Airways' CPTP or WTS series. In the four years of the CPTP and WTS courses, Kent State met 97 percent of its allotted quota.[26] The end of the WTS also marked the end of Kent State's direct involvement with flight training until after the war.

Kent Flying Club

Kent Flying Club was still active during this time. Since the club's formation in 1939, it had successfully operated out of Stow Field until sometime in 1941. Deteriorating runway and facility conditions most likely prompted the club to move to Mid-City Airport. Once the club left Stow Field, there was little else happening there. When private flying resumed after the Pearl Harbor attack, club members initiated a rule that for every takeoff made, a portion of the airplane rental fee would be used to buy a defense stamp. The flights were tallied at the end of the month, and defense stamps were purchased.[27]

In the summer of 1942, a student pilot with Kent Flying Club, who had yet to earn his license, took a friend for a ride, which was against CAA regulations. Soon after taking off from Mid-City, the airplane crashed, killing both the pilot and his passenger. Two weeks later, the president of another club crashed his airplane into utility lines near Green, Ohio. While there were no injuries in that incident, two accidents in such a short period drew attention from the CAA.[28] Immediately following the second accident, the CAA revoked the permits for eight flying clubs in the area, affecting one hundred pilots. The pilots' licenses were temporarily suspended.[29]

The last mention of Kent Flying Club came in August 1942, when its newly purchased airplane was damaged by hail at the Sky Haven Airport in Northfield.[30] Losing favor with Mid-City management likely had prompted the relocation of the club to Sky Haven. Buying a new airplane to replace the one lost in the fatal accident, and then having it damaged so soon, might have been the undoing for Kent Flying Club.

Private Flying Impacted

Further pressure on the private flying community was announced in July 1942, when flying would soon be prohibited at Akron Airport. In December 1941, the Goodyear Aircraft Corporation had been awarded a contract to build the Vought F4U Corsair fighter at the Akron Airport, requiring that all private flying be forbidden. The first flight of a Goodyear FG-1 Corsair did not happen until February 1943, but by then one hundred private pilots at the Akron Airport had been forced to relocate. Since CPTP pilots were government-sponsored, they were still given access to Akron Airport as it was the only local airport lighted for night flying.[31]

Stow Field was another possible choice for private flying, but the field was poorly maintained, and the hangar was unusable. Ray Van Devere, Rudy's brother, stated the airport was too expensive for the private pilots to lease and operate. "We have not made up our minds yet what to do with it."[32]

Possible Airport Sites Explored

The WTS program expansion in 1942 provided another incentive for Kent State University to locate and purchase an airport. A year had already passed since the state had authorized the money to acquire an airport, yet the search continued. Stow Field seemingly received only a passing consideration,

One of the FG-1 Corsairs built by Goodyear Aircraft Corporation at Plant D, Akron Municipal Airport, 1943. Andrew Paton wrote technical manuals for the Corsair while employed with Goodyear during the war. (William W. Schloman Jr. Collection)

possibly because it had seen little use over the past few years and was in poor condition. However, given the CAA requirement that any airport used for the CPTP needed to be within ten miles of the university, options were limited.[33]

In June, KSU President Karl Leebrick announced that Kent State would select a site for the proposed airport by the middle of August. Eight sites were under consideration, including Stow Field, Mid-City Airport, and land in Kent, Twinsburg, Brimfield, and Brady Lake. This likely prompted Rudy Van Devere to announce he was fixing up Stow Field to be used by CPTP students. Final selection hinged on the cost of the land and needed grading, possibility of future expansion, and proximity to the university.[34]

Kent State University Purchases Stow Field

On September 11, 1942, the board of trustees requested $50,000 from the Ohio Board of Control for the "purchase of the necessary land for the development of the Stow Airport." An announcement was made on October 1 that the state had released $46,500 to Kent State for the purchase and development of the 198-acre Stow Field. Sixty-five percent of the amount was for the actual purchase and the rest for improvement of the site.[35]

Notably, the purchase of the airport by the university did not include the parcel with the hangar or airport buildings. It is not clear why the buildings were excluded from the sale. Was it a university decision or a desire by Van Devere to retain a presence at the airport? Whatever the reason, split ownership would create tension and problems in the years ahead. The *Akron Beacon Journal* reported that Stow Field was to be renamed "Kent Airport," but any mention following the sale used the name Kent State Airport. Van Wyen expected that the airport would be ready for student pilot training by May 1943. Until then, KSU students would continue to train at Mid-City Airport.[36]

On October 5, 1942, Kent State officially took over Stow Field and continued the grading work that Rudy Van Devere had begun. Although the university had bought 197 acres, not all of it was to be developed. According to the selection criteria, the university would set aside the unused acres for later expansion when needed.[37]

Elsie and Rudy Van Devere signed over the necessary deeds in late October 1942 from their residence in Helena, Arkansas, where Van Devere had established another CPTP flight school. With receipt of the paperwork in November, the university promptly sent it to Columbus to obtain the title from the state.[38] It had been nearly a year and a half since the state had given its initial approval for purchase of an airport. At last, Kent State owned an airport.

336th College Training Detachment

In early 1943, as the WTS was about to end, Kent State received two inquiries from government agencies about using university facilities. The first asked to establish a high school teacher training course in preflight aeronautics. President Leebrick responded that the university would participate in the program. The training course was available to a limited number of high school teachers and covered general service and operation of aircraft, meteorology, navigation, and civil air regulations. Participation was limited to those currently teaching or to those who could certify they intended to teach preflight aeronautics in a secondary school. Andrew Paton, Edward Curry, and Adrian Van Wyen were selected to be the instructors for the KSU course.[39]

The second, and more significant, inquiry was from the War Department, requesting that Kent State consider participating in an Army Air Forces College Training Program that would send cadets to universities where CPTP facilities existed. Upon receiving the telegram, Leebrick had until midnight the following day to respond. He quickly consulted the faculty and received its unanimous support. The board of trustees telegraphed a response indicating Kent State's desire to participate.[40]

On March 30, 1943, five hundred army air force cadets arrived in Kent via train as the newly formed 336th College Training Detachment at Kent State. They were organized into five squadrons of one hundred cadets each. Each squadron had three sections (or flights) of thirty-three aircrew students with four squads to a flight. Dormitories were converted into barracks, and cafeterias became mess halls. The planned graduation rate was for one hundred cadets after two months, and one hundred cadets each month following. The detachment's strength was maintained with an additional one hundred cadets arriving each month.[41]

Enrollment at Kent State had slumped from 2,700 prewar to 600 during the war. The arrival of the 336th filled and enlivened a school that was struggling. William H. Hildebrand, in his history of the university, maintained that, given the low enrollment and a payroll with too many faculty, "had it not been for the 336th Training Detachment's presence, KSU would probably have ceased to exist before war's end." Approximately 1,950 enlisted men would complete the program at Kent State.[42]

336th College Training Detachment cadets arrive by train for a five-month training course, March 30, 1943 (William D. Schloman Collection)

Classes began on April 5. The cadets took classes in mathematics, physics, history, geography, English, civil air regulations, and medical aid. Kent State also provided 80 hours of supervised study and 120 hours of physical training for each cadet. Most of the coursework was already offered at the university, but now there was more focus on aeronautical themes: mathematics for aerial navigation, geography in aeronautical charts, and aircraft ignition systems in physics. Andrew Paton left his previous role as a CPTP instructor to teach physics.[43]

Some of the university departments adopted special textbooks for their military courses. This was to keep the subject matter applicable to their training, and also because the regular textbooks proved too difficult for the cadets. Faculty members found that the cadets did not achieve the same level of mastery of subject matter as expected in civilian classes. Heavy schedules and accelerated timelines during their stay at Kent State did not help. More importantly, there was no requirement to be a high school graduate for admission into this College Training Program, putting many participants at a significant disadvantage.[44]

College Training Detachment cadets marching up Summit Street in Kent, singing Air Force songs, March 31, 1943 (Kent State University Libraries. Special Collections and Archives)

Flight training was handled quite differently than during the CPTP years. All aspects of the training, both ground and flight, were contracted to Akron Airways and under direct supervision of the army. Kent State had no connection or responsibility with anything regarding flight training. The last month of a cadet's training was called "Flight Alert," during which the cadet was given ten hours of dual flight instruction. This training was similar to the first block of training provided to CPTP students. While the goal for the CPTP was to have the student solo the airplane and progress toward a private pilot license, the army training consisted entirely of dual instruction.[45]

The Kent State University Airport was not ready for use by the cadets by May 1, 1943, so they began their flight training at Mid-City Airport. Ray Van Devere's distinctive orange bus provided transportation for the cadets between the campus and Mid-City. One of the instructors teaching these young men to fly was Richard Schwabe, who later became an important figure in Kent State's flying program as instructor and mentor.[46]

As 1944 started, the board of trustees authorized the university to negotiate with the army to "get more Aircrew students for the University." As a result, the university's quota was increased to 570, and the monthly graduation and replacement rate increased to 114 cadets. While the plan was to maintain the

Top: Mid-City Airport with line of Piper J-3 Cubs, 1943 (Van Devere Family Collection)

Bottom: 336th College Training Detachment sign for restricted area on campus, 1944 (Kent State University Libraries. Special Collections and Archives)

initial level of 500 cadets, then increase it to 570, the actual numbers never quite materialized. Within five months after the arrival of the first 500 cadets to Kent State, their ranks dropped to 200. The problem was that the army insisted on an outflow of one hundred cadets every month, but they did not replace them with an equivalent number. The required outflow also meant that frequently the cadets' five months of training were not completed. Approximately 1,200 cadets participated in the program at Kent State, yet only 200 actually received the full amount of planned training. Courses were reduced to include only the most important parts of physics, math, and geography. The only course that was provided as planned for all cadets was flight training.[47]

During the federal flight-training years at Kent State, only one fatal accident occurred. On March 2, 1944, instructors Joseph Smith and Robert Long departed from Mid-City Airport in their Piper Cubs with students Myron Spencer and Norman Zidar. The two airplanes flew southwest and entered a forty-square-mile practice area along with two other training airplanes over Steels Corners Road in western Stow. While performing low-level maneuvers, the two airplanes collided, killing the two cadets and two civilian instructors.[48]

End of the College Training Program

In April 1944, the War Department announced that the Army Air Forces College Training Program would terminate by June 30 due to an oversupply of trained cadets. This decision triggered a name change for Kent State's unit.

The 336th College Training Detachment was inactivated in May and then activated as the 2586th Army Air Forces Base Unit. Despite the impending end to the program, the army shipped an additional 243 cadets to Kent State. The program cancellation affected 400 cadets there. Nationwide, there were 120,000 cadets at eighty-one colleges and universities participating in War Department programs. Plans were to transfer 40,000 students a month into on-the-job training at airfields in the Army Air Forces Training Command. Flight-training operations at the Akron Airport ceased, while those at Mid-City continued until the last students shipped out on June 28.[49]

As classes ended, the university had to reduce staff. Andrew Paton submitted his resignation in the middle of April and returned to work at the Goodyear Aircraft Corporation as a maintenance engineer.[50]

Adrian Van Wyen at his desk in the Pentagon, 1946 (Adrian Van Wyen Collection)

Adrian Van Wyen left in August 1944 to join the US Naval Reserve as a lieutenant (junior grade). At the end of the war, Van Wyen accepted a position from the navy to stay on in the history unit. By 1961, he had published several books and was chief of the aviation history unit in the office of the Deputy Chief of Naval Operations for Air.[51] He had served the university well from 1939 to 1944 in directing the several federal programs that had provided training to hundreds.

Kent State's participation in the College Training Program was not as advantageous for future flight training as it had first appeared. The university's enthusiasm to participate reflected the school's desire to cooperate to their fullest extent with the war effort. With enrollment suffering as the war progressed, it became evident that the number of faculty exceeded the needs at the time. Although CPTP participation met the assigned quotas, those quotas were continuing to decrease. The influx of hundreds of army cadets and their educational program had provided much-needed work to Kent State.[52]

Even prior to adopting the College Training Program, Kent State had found the army's track record with planning to be poor. In both the WTS and College Training Program courses, the army showed great interest in producing trained aviation cadets yet had little plans for what to do with them once they finished their coursework. The College Training Program produced trainees in numbers that exceeded the army's need for pilots and even aircrewmen in general. The haphazard college training each cadet received had little value and was seen as a waste of effort and resources. The university itself saw no lasting benefit from its involvement.[53]

Years later, Paton reflected that a better path for aviation at Kent State probably would have been to stay with the previous WTS program for naval aviators. The naval air corps recognized the training received in the WTS as an equivalent to their own primary training. WTS graduates seamlessly moved into advanced naval air corps training stations upon graduation. The naval programs were never expanded like those of the army, which created a backlog of cadets. After the war, those schools that had stayed with their WTS training programs frequently continued with naval ROTC programs. Maintaining a military aviation unit at a university benefited a school's aeronautical program and its physical facilities. Reportedly, Bowling Green State University built new facilities for its program from the revenue it earned from its WTS and other naval programs.[54]

Moving Forward

Even with a world war raging on two fronts, the university was making plans for a future aviation program and use of its airport. In November 1943, Van Wyen and Paton were named to the university's War Activities Council to investigate possible postwar courses in aeronautics. The university also began developing plans to request State funds for a postwar building program, including a request for its own airport shop and hangar. As with many other airport building requests that would be made over the next decades, nothing came of this plan.[55]

With the announced end of the College Training Program in 1944, the university took advantage of an opportunity to enter into a contract with the Firestone Tire and Rubber Company to provide ground school instruction to be held at their facility in Akron. Flight training and equipment were provided by Akron Airways. In June 1944, the Servicemen's Readjustment Act, also known as the GI Bill, became law.[56] As a result, universities expected a resurgence of postwar enrollments. Given the success with both the CPTP and WTS, they hoped the same could be expected after the war ended.

Wartime Experiences of CPTP Graduates

Throughout the country, CPTP graduates began their careers in wartime. The stories of some CPTP KSU graduates are representative of what many of their peers faced and what some sacrificed.

Henry Russell Kenyon (1916–1942) was one of Kent State's first CPTP graduates, leaving in November 1940 to enlist in the navy. Trained as a pilot on the Douglas TBD Devastator torpedo bomber, Kenyon joined Torpedo Squadron Eight (VT-8), which was assigned to the aircraft carrier USS *Hornet.* On May 28, 1942, the USS *Hornet* sailed from Pearl Harbor with two other carriers to repulse an expected attack on Midway Island. On June 4, Ensign Kenyon launched from the deck of the USS *Hornet* with the rest of VT-8 seeking to knock out Japanese aircraft carriers. This was the first time the pilots had taken off from an aircraft carrier or even carried a torpedo. When the group's fighter escort took an incorrect heading, the leader of VT-8 broke formation and followed the correct heading. The error resulted in inadequate fuel for the pilots to complete the mission and then return to their carriers. The fighter aircraft had to turn back or ditch into the ocean, leaving the fifteen aircraft of VT-8 unprotected. The slow, torpedo-ladened aircraft were no match for the Japanese fighters, and one by one the Devastators fell from the sky. There was only one survivor out of the crew of thirty. Kenyon perished and was awarded the Navy Cross and Purple Heart. Over a year later, Kenyon's widow received notice that a navy ship would be named for her fallen husband. The USS *Kenyon* went on to serve as an escort for convoys in the Caribbean, Atlantic, and Mediterranean before finishing out the war serving in the South Pacific.[57]

Top: Ensign Henry Russell Kenyon (William D. Schloman Collection, Donated by Russellyn K. Edwards)

Bottom: Capt. Jack Wardell Petit (Ann Hendricks Collection)

Jack Wardell Petit (1921–1943) joined Kent State's inaugural CPTP class and became the first in Ohio to complete the secondary course. Petit enlisted on October 1, 1941, and began his training as a US Marine Corps fighter pilot. He joined his squadron, the VMF-214 "Swashbucklers," in Hawaii in 1942. Flying Grumman F4F Wildcats, the pilots flew missions to protect marines fighting on the ground at Guadalcanal. This first combat tour ended with numerous victories and no pilot lost. For his second tour, the squadron flew the powerful Vought F4U Corsair. Petit and two other Corsairs attacked three Japanese bombers that were striking at a US destroyer towing a damaged seaplane tender. Each pilot took down a bomber without a scratch. Petit's third tour was with the VMF-215, "The Fighting Corsairs." Tasked with covering the US Marine invasion of Bougainville Island ninety miles from their base at Vella Lavella in the Solomon Islands, the pilots flew four-hour patrols over Bougainville from dawn to dusk. On October 20, 1943, Petit and the two other Corsairs flew in a *V* formation to return to their base while maintaining radio silence. Adverse weather separated the airplanes, and only two returned to the base. Rescue flights found no trace of 1st Lt. Petit or his Corsair. With less than a month left in his third and last combat tour before returning to the United States, Petit lost his life in the skies over the South Pacific.[58]

Mary Sherriff (1920–2014) was one of four women accepted into Kent State's first CPTP class and the first among those to solo. Women in aviation were an exception at the time and were not seen as equals by their fellow students or instructors. Sherriff's strong work ethic was reflected in her piloting ability, causing even those with bias against female pilots to see her as exceptional. When her flight instructor noticed Sherriff repeatedly making the same mistake, he had her write "Use throttle slowly and smoothly" on the blackboard 200 times. Success followed, and she earned her private pilot license in April 1940, the first woman in Ohio to do so as part of the CPTP. In a newspaper article highlighting her achievement, Sherriff stated her wish to become a stewardess with Pennsylvania Central Airlines, as they were the only airline not requiring female applicants also to be nurses. The airline's president saw the article and immediately offered her an interview and then a job. Leaving in July 1940, Sherriff became the first KSU student to gain employment because of the CPTP. Sherriff's career as a flight attendant ended with the decision to start a family, but she still inspired her granddaughter to become a pilot. That young woman would eventually work for United Airlines, which started out as Pennsylvania Central Airlines—the same company through mergers that her grandmother worked for seventy years earlier.[59]

Mary Sherriff, stewardess with Pennsylvania Central Airlines (Stubbs Family Collection)

Lowell "Buzz" L. Wetzel (1920–1943) came to Kent State in 1938 along with his friend Richard "Dick" Schwabe. He was accepted into the CPTP program in January 1941, earning his pilot license in April. Following the attack on Pearl Harbor, Wetzel enlisted in the US Army Air Forces and joined the 331st Bomber Squadron, 94th Bomb Group as a B-17 Flying Fortress pilot. Following months of training in Florida, Arizona, and Kansas, he and his crew were assigned to a brand-new B-17 bomber, which they named *In Der Fuehrer's Face* and had a skunk painted on the nose. On the first leg of their journey to Europe, Wetzel flew over Kent and buzzed his parents' home and his new wife's sorority house (later learning that they were going to be parents). Once in England, Wetzel and his crew began flying missions over Europe. Engine reliability was an ongoing problem, so most missions were aborted before reaching the target. On May 21, 1943, the 94th Bomb Group was assigned a mission to bomb the U-boat shipyards in Emden, Germany.

In Der Fuehrer's Face was given the difficult task of being "Tail End Charlie" for this mission—that is, the last bomber in the last squadron in the last group at the lowest altitude. The intense forty-five-minute engagement with German fighters ended when shells from a Messerschmitt Bf-109 detonated the B-17's bombs and obliterated the bomber over the Holland countryside. Less than three months before his twenty-first birthday and on his third mission, Buzz Wetzel lost his life in the skies over Europe. His son would also have the nickname Buzz and develop strong connections to Kent State.[60]

Top: 2nd Lt. Lowell "Buzz" Wetzel and wife Dorothy (Lowell "Buzz" Wetzel Starner Collection)

Bottom: Wayne Wilkin flying for the Air Transport Command (Wilkin Family Collection)

Wayne Wendell Wilkin (1917–1995) was in Kent State's inaugural CPTP class. Offered a job from United Airlines, Wilkin left Kent State with less than one year needed to graduate and within two years became a United Airlines captain, flying the San Francisco-Denver route. He took a position with American Airlines in 1944 as a captain flying its transatlantic routes. The airline had contracted with the army air force to transport cargo to Europe and return wounded soldiers. The emotion from his fifty trips of bringing the wounded home remained with him his entire life. In 1945, pilots were in short supply in Europe, leading Wilkin to join KLM Royal Dutch Airlines as a captain. While flying its transatlantic routes, he arranged with both the airline and Kent State to have a schedule that would allow him to attend classes and complete his degree. Wilkin graduated from Kent State in December 1948. When he retired from KLM in 1973, he had made over 1,000 flights across the Atlantic, flown numerous aircraft including the Boeing 747, and recorded 23,741 hours in his logbook. Wilkin passed away on April 14, 1995.[61]

Summary

The United States' entry into World War II brought many challenges to Kent State University and its aviation education. Shifting from a primarily civilian program to one meeting the needs of the war effort appeared seamless, but there is little doubt that it was due to the leadership of Adrian Van Wyen and Andrew Paton. Their success gave the university confidence that buying an airport was an investment in a future with collegiate aviation once the war was over. Paton left the university when training programs ended, but his expertise would be needed when the university started its own flight-training program in early 1946.

Chapter 7

Establishing Aviation Education, 1945–49

Kent State University saw a change in leadership when George A. Bowman became its fifth president on July 1, 1944, the day after the 336th College Training Detachment left campus.[1] Many postwar issues would face Bowman, including those related to managing an airport. Runway conditions were troubling, and there was a lack of equipment and men familiar with doing the needed work. The biggest undertaking for Kent State as the war concluded was to institute a flight-training curriculum at the university. The Civilian Pilot Training Program (CPTP) years had seen success, and now as airport owners the university had to take advantage of that expertise.

Aviation in Ohio had seen a boon of activity during the war years; however, there were only fifty airports in twenty-two northern Ohio counties, with several counties having none. In December 1944, the Civil Aeronautics Authority (CAA) recommended to Congress that $31 million be allotted to Ohio for the construction of 129 new airports and improvements to another 77. Unfortunately, Kent State was omitted from the official congressional report regarding airport improvements. It was discovered that the university never filed the required CAA paperwork for change of name or ownership after purchasing Stow Field. The congressional report did provide for Stow Field to be developed as a Class 2 site at a cost of $64,000. The problem was that Stow Field as an airport name did not exist anymore. Bowman communicated with the director of airports at the CAA to have the issue resolved. By the end of 1944, the CAA was informed that the Kent State University Airport was an operational facility and should be included in all future records.[2]

Finally, on September 7, 1945, Kent State completed the paperwork to the CAA designating the airport as Kent State University Airport. Lawrence Wooddell, KSU superintendent of maintenance, was selected as the first manager for the airport. Wooddell would go on to make KSU history by introducing black squirrels into northeast Ohio and specifically the KSU campus. The black squirrel would became an unofficial mascot of the university.[3]

Filing the needed CAA paperwork led to an official CAA identifier for the airport. The Civil Aeronautics Act of 1938 introduced a three-letter airport identification system for airports with control towers or navigation aids, scheduled air carrier, military airlift service, or designation as a port of entry. Public-use airports that did not meet the requirements for a three-letter identifier, such as the KSU Airport, were given a one-letter, two-number identifier. The KSU Airport received the identifier of 1G3.[4]

Airport Improvements Needed

Work began by the end of 1944 to tackle the ever-present erosion problem at the airport. Seeding was completed and logs were placed along the tree line to combat the erosion, but it was not enough. Rudy Van Devere proposed that in exchange for fixing the runways, he be given the privilege of occasional landings at the field. Bowman approved the arrangement. By March 1945, the university was displeased with Van Devere's progress. Although Van Devere had the equipment, he lacked the manpower to run it. He promised to bring help upon his return from a trip to Louisiana to do significant work at the airport. The arrangement at the time—Kent State owned the airport property while Akron Airways, Inc., owned the facilities and provided the services—would become more and more of an issue as the months and years wore on. Tensions would continue to rise between the parties.[5]

Van Devere now envisioned a business venture at KSU Airport, and to that end he built a 3,000-square-foot hangar of cement block in 1945, replacing the dilapidated hangar he found when he purchased the property in 1941. His new hangar would become known as the maintenance hangar and possibly was built on the footprint of Hugh Robbins's earlier hangar. Adjoining the hangar was a 1,520-square-foot cement-block building used as a shop and office. In the rear were a small-frame barn and storage shed.[6]

In August 1945, Van Devere made a request to President Bowman for exclusive commercial operating rights at the airport, both flight instruction and charter flights, in exchange for providing ground maintenance. He offered his maintenance services at sixty to seventy dollars per month, while

First hangar built by Rudy Van Devere at the Kent State University Airport, 1945. It would become known in later years as the maintenance hangar and possibly is built on the footprint of 1929 Hugh Robbins's 1929 hangar. (William Andrew Paton Family Collection)

he would pay roughly the same fee for the commercial operating rights. It would be a break-even proposition for Kent State and would ensure a future for Akron Airways at the airport. With discussion about starting a flight-training course like the CPTP after the war, this partnership probably seemed to the university to be an easy and proven solution. The university required that Akron Airways furnish all necessary equipment for airport operations; act as an assistant manager under Wooddell; maintain the airport, including mowing, fertilizing, and reseeding; and provide a 10 percent discount on all flight training for KSU students. For the next two decades this arrangement, born during the war, would continue. In October 1945, Van Devere bought the large Quonset-style hangar that would become the iconic identifier for the airport in the years to come. It became usable by the summer of 1946.[7]

By the fall of 1945, the airport was still in need of maintenance. The university was requiring that all flight training take place at its airport, in contrast to the earlier federal programs. Van Devere presented the university with a list of airport deficiencies, including lack of a hangar for airplane storage and poor drainage that caused a soft field. The airport had to be approved by the VA (Veterans Administration) in order to qualify for training under the GI Bill (Servicemen's Readjustment Act of 1944), leading to concerns that these issues could impact that decision.[8]

The university lacked funds to address the hangar and drainage issues. Van Devere's stopgap solution to the storage problem was to set up storage hangars next to the one hangar he had already constructed. A temporary fix for the field conditions was to close the south end of the north-south runway until a solution was found. Before the year was out, the KSU Airport did see some activity when Akron Airways suspended its operations at the Akron Airport and sent students to Mid-City and KSU Airports. Under snowy conditions, it was easier to land an airplane equipped with skis on a field than on a paved runway.[9]

Kent State University Flight Curriculum Developed

With the conclusion of World War II in early September 1945, Kent State called Andrew Paton from his wartime employment with Goodyear Aircraft Corporation to develop an aeronautics curriculum, including flight training. The GI Bill provided money to returning servicemen to attend college, and the university wanted to position itself to attract these young men. To do this, the university had to set up the courses, establish their costs, and have them approved by the VA. The goal was to have the program up and running by the beginning of January 1946, in time for winter quarter.[10]

By November 1945, Paton had put together proposals for the new flight courses. The elementary flight course would be fifty-six hours of flying in light airplanes, plus one hour for the student's final CAA flight examination. The flight-instructor course would be fifteen hours in light airplanes and five hours in the larger Fairchild PT-19 airplane, plus one hour for the final CAA flight examination. The flight-instructor course was aimed at pilots who already had their commercial certificate and wanted to get their instructor certificate. Flights would be scheduled during school hours, as well as at night and on weekends. The university would pay Akron Airways nine dollars per hour for light airplanes and fifteen dollars per hour for the PT-19 airplane, both for solo and dual instruction flights. Akron Airways would additionally provide a 10 percent discount for these flights.[11]

All six new aeronautics courses were approved in December, including two ground school classes and the two flight-training classes. The approved courses provided training in three distinct areas: general aeronautics, mechanic training, and flight training. Flight training was to be offered for the sole purpose of expanding the education of regular students at the university. Any training beyond what the university planned to offer would be the function of an established flight school, and not Kent State. Kent State was an institution of training teachers, so flight training offered beyond the initial private pilot certificate was to bridge the gap to the flight-instructor course.[12]

Students started enrolling for the courses as soon as they became available for winter quarter. Initially, the flight-training courses were available only to army air service veterans currently enrolled at Kent State. Although they would be available to other college students in the future, only full-time students would be eligible. Paton called these courses "a privilege offered to regular students."[13]

The KSU Airport, as well as Akron and Mid-City Airports, received the needed approval to provide GI flight training at the beginning of 1946. Akron Airways added three new flight instructors returning from the army. One of

Andrew Paton takes student reporter Jean Olson through preflight checks in a Stearman PT-17 Kaydet, the *Daily Kent Stater,* January 22, 1946 (Kent State University Libraries. Special Collections and Archives)

these was Richard Schwabe, who would become a significant part of flight training at Kent State later in its history. When classes began in mid-January, there were seventeen veterans enrolled in flight training. One was Kent Eby, who had been an aviation cadet in Kent State's 336th College Training Detachment. He went on to serve as a gunner on a B-24 Liberator bomber and returned to Kent State to follow up on his desire to become a pilot.[14]

Flight training had a small setback in February 1946, when 68 mph winds hit the region and tore from their moorings two of the four Akron Airways' aircraft being used for training, damaging them beyond repair. Other than this episode, the quarter finished without any other issues. The university extended its agreement with Akron Airways, granting exclusive commercial operational rights at the airport. Kent State had the right to cancel the agreement with twenty-four-hour notice, but that was a moot point. Given that Van Devere's two-acre property was adjacent to that of the university's airport, Kent State had little recourse to stop him from doing what he wanted should they terminate the agreement.[15] This was one more reason they hoped to purchase his property at some point in the future.

Above: Mid-City Airport aerial view, looking north toward Hudson, Ohio, 1946 (Stanley A. Crosier Collection)

Right: Mid-City Airport hangar with Rudy Van Devere's business, 1946 (Van Devere Family Collection)

Flight Training with the GI Bill

Funding by the GI Bill proved to be a boon to flight training in spring quarter. Kent State was forced to limit enrollment to what six flight instructors could handle, creating a backlog of interested applicants. This had a psychological effect on students. Knowing others were waiting, students were careful not to miss a scheduled flight time and did not object to flying during the winter. Kent State ended up with forty-one students enrolled in flight training, including six women—five of whom were veterans. Akron Airways added two Piper Cubs to the fleet at Kent State, bringing the total to five.[16]

With the increased interest, Kent State boasted more flight students than Ohio State, and the faculty claimed that they were the only college in the state with an airport so close to campus. However, little had been done to recognize Kent State's ownership of the airport outside of aviation circles. One anonymous faculty member wondered why Van Devere did not "stick up a big sign proclaiming this as 'Kent State Airport.' No one could tell from the road what field it is."[17] One might also ask why the university did not erect its own sign for the airport.

Instructor with student in a Piper J-3 Cub at the KSU Airport, 1946 (Kent State University Libraries. Special Collections and Archives)

Increased demand by veterans for flight training meant increased revenue for the university. The GI Bill gave each veteran $500 per year for college. At the time, tuition at Kent State was $125 to $150 per year, leaving enough money for flight training. The private pilot course cost $364, and the flight-instructor course was $225. This income was a lifesaver for schools on the verge of bankruptcy at the close of World War II. Nonetheless, the success of the GI Bill in bringing new students to college put a huge burden on resources at tax-supported schools. In order to limit how many classes a veteran could take, the VA on May 20, 1946, authorized institutions to charge each veteran $35 per term or $135 per year more than nonveteran students. This virtually doubled a veteran's education bill and made paying for flight training impossible.[18]

To stretch a veteran's GI Bill financial aid, Kent State split the private pilot course into two courses.[19] A veteran could now spread flight training over two school years and have enough money left over in each year to take flight training, even with the extra fees imposed by the VA. In addition to raising fees at the university, the GI Bill had the effect of raising rates among flight contractors.[20] The VA also established maximum rates that contactors could charge for flight training. As a result, all contractors raised their rates to the maximum, prompting nonveterans to find cheaper options elsewhere.[21]

Paton saw the revenue from flight training providing significant profit for Akron Airways, prompting him to write Dewey Barich, head of KSU's industrial arts department: "I sincerely hope that it will soon be possible to divert this sizeable fund into our University Treasury for the purpose of our own flight-training program. There is no question in my mind but what these rates could be strongly reduced and still leave the University a margin by means of which the aeronautics program could be built up and equipped." When shared with President Bowman, he responded that he was not as optimistic as Barich and Paton "regarding the desirability of the University getting into the business of operating an airport. I hope you are taking into consideration all of the implications of such a move."[22] Paton's vision for flight training and use of the airport would remain constant for decades to come.

A new piece of war surplus equipment arrived at Kent State with delivery of a Link Trainer from Greenville, Mississippi, in May 1946. The flight simulator was used to train military pilots to fly by instruments only. It would be used in the aerodynamics course, with the hope that it could lead to the development of other flight-training courses.[23]

A cross-country course was added to the flight-training curriculum in spring 1947. Although the CAA did not require a cross-country course, Akron Airways and Kent State partnered to offer one. "Too many young

Andrew Paton with student in the Link Trainer, a surplus World War II flight simulator, 1946 (Kent State University Libraries. Special Collections and Archives)

pilots go out and get lost—run out of gas eventually and crack up," explained Rudy Van Devere. "In order to forestall accidents with our college groups we are adding this new course. I believe it gradually will be copied and may become a government requirement." The course would include twenty-six hours of cross-country flying in a Stinson 150. Each flight would be a total of 400 miles in length and accomplished via dead reckoning (a navigation method using known waypoints and observed wind effects to calculate course corrections en route).[24]

The growth of flight training at the KSU Airport also brought increased complaints from local residents not keen on airplanes flying overhead. Paton fielded these complaints, usually brought about when someone thought an airplane had "buzzed" a part of town. Being the biggest operation in the area

Stearman with its fabric removed in the aeronautics shop in a temporary building on campus, 1946 (Kent State University Libraries. Special Collections and Archives)

meant the blame was usually placed directly on university operations at the airport. The KSU Airport was also a public facility, so there was plenty of airplane traffic not associated with university-sponsored flight training. Paton followed up on each complaint and routinely traced the offending airplane to those flown out of other nearby airports or individuals who were not KSU students.[25] This was an early introduction for the university as an airport owner in dealing with public concerns, a community relations task that would be important and ongoing.

The KSU Airport had never seen flight training at this level. Nearly all of Kent State's wartime flight training had taken place at either Akron or Mid-City Airports. Now that Kent State had established its own flight-training courses, it made every effort to ensure that all instruction happened at its airport. Wet and snowy conditions made the airport muddy, and in the summer airport operations included dealing with blowing dust. Van Devere gave Wooddell a list of needed university improvements, including paving a portion of the parking area and a strip to and from the large hangar. Akron Airways would also pave a large area in front of the hangar. President Bowman reported to the board of trustees that state appropriations would be inadequate to make any of the suggested improvements.[26]

Hangars signed with Kent State Aircraft, Rudy Van Devere's business offering flight training under the GI Bill, 1948 (Kent State University Libraries. Special Collections and Archives)

The changes with the GI Bill financing started to be felt by Akron Airways by late summer, causing Van Devere to approach Kent State about raising flight-instruction rates. He also informed the university that he could no longer operate as an approved CAA school under the Akron Airways name. He received approval to provide GI Bill training under the name Kent State Aircraft. Bowman informed Van Devere that he had appointed Andrew Paton as airport manager beginning with fall quarter 1946.[27]

Flight training continued to flourish at Kent State. By the end of 1946, 125 students were registered for the following winter quarter, thanks in part to nonveteran students being allowed to enroll. In addition, there were sixty-five students enrolled for ground school with anticipation of that number doubling by the time school started. To handle the increased number of students, a second elementary flight course was added. Van Devere doubled the number of flight instructors at all three airports to meet demand and added new all-metal Cessna airplanes to his fleet.[28]

Other flight-training contractors were undoubtedly aware of the success of flight training at Kent State. In an effort to compete for that business, Horn's Flying Service approached Kent State at the start of winter quarter 1947 to take over flight-training operations at the airport. While there was initial interest by Kent State, the offer was not pursued. While nothing came of this proposal, it would not be the last time the university contemplated other flight-training arrangements.[29]

The Kent State University Flying Club

In spring quarter 1947, flight students expressed an interest in forming a flying club. The National Intercollegiate Flying Clubs (NIFC), which had become inactive in 1942, announced it was reorganizing. In addition to bringing flight students together, it fostered an interest in light airplane flying and safety. NIFC was the predecessor to the present-day National Intercollegiate Flying Association (NIFA), which holds regional and national competitions for university-sponsored flight teams.[30]

The primary objective for creating a KSU NIFC chapter was to offer economical flight time to flight students. The biggest hurdle was to meet the NIFC requirement for the club to have an airplane. Since all Kent State flight training was provided by Akron Airways, the university did not own any airplanes for the students to use. In July 1947, the Kent State University Flying Club organized with Andrew Paton as advisor. Shortly thereafter, the club met with a Cleveland airplane broker to talk about buying and operating its own airplane.[31] It took until December 1947 for the KSU Flying Club to buy its first airplane, a 1941 Luscombe Silvaire. To raise the necessary money, the club had incorporated that October. Members were required to pay for a share of the airplane, which then gave them rental rights.[32]

Setbacks plagued the KSU Flying Club in 1948. In late February, only two months after buying its first airplane, the Luscombe Silvaire was damaged beyond repair in an accident at the airport. Two months later, the club bought a replacement airplane, an Aeronca Chief with 160 hours of accumulated time. Just a month later, a member spun the aircraft while turning into a final approach at the KSU Airport, severely injuring himself and his passenger. A second replacement airplane, a 1940 Piper Cruiser, was purchased ten months later.[33]

Piper Cruiser being disassembled for transport to the industrial arts shop on campus to do required maintenance, 1949 (William Andrew Paton Family Collection)

Even with these newsworthy accidents, the KSU Flying Club prospered. With its own airplane, the club was now eligible to participate in NIFC meets. To prepare for the national meet to be held in Detroit the next year, a flight team was assembled from private and commercial pilots currently registered at Kent State. There was no requirement to be either a KSU Flying Club member or enrolled in an aeronautics course. Pilots trying out paid a portion of the flight costs with the university covering the rest. The university also covered the complete cost of the airplane during meets. A six-member team was selected to represent Kent State at the upcoming competitions. Paton saw this

The Kent State University Flying Club with their 1941 Luscombe Silvaire, 1948. Front row (*left to right*): George Williams, William Anderson, Donald Wohlford, John Lilley, Jim Butcher, and Richard Johnson. Back row (*left to right*): Frank Kovasic, Prof. Luey, John Lyon, George Ebel, Harold Province, Bernard Zents, Ronald Armitage, Edward Mayer, Tad Stokes, and Prof. Andrew Paton. (Kent State University Libraries. Special Collections and Archives)

as an opportunity to create interest in college aeronautics and to offer student pilots precision flying experience. To prepare for the national competition, plans were made to hold meets with other colleges and universities.[34]

By the end of 1948, the KSU Flying Club had grown to twenty-five full members and ten associate members. A full member had paid for a share in the club's airplane and had full rental rights. Associate members could not rent the airplane, but they were able to participate in club activities and go on flights with other members. Once there were ten more full members, the club planned to purchase a second airplane.[35]

On May 19, 1949, the first meet for the KSU flight team was held at the KSU Airport against four pilots from Horn's Flying School of Chagrin Falls. Coached by mathematics instructor William C. Lowery, the team competed in four events: bomb drop, precision landings, low-altitude pattern flying, and navigational searches. Kent State won this first flight team competition and quickly set its eyes on an upcoming meet at Kalamazoo, Michigan.[36]

Changes to GI Bill Benefits

Late summer 1947 brought a significant change to GI Bill benefits for veterans using their extra allotted education funds to pay for flight training. The VA deemed that flight training was not required for a college education. To penalize those veterans who had used their funds toward flight training, the VA reduced their period of entitlement one day for every $2.10 spent. Upon leaving the armed forces, a veteran was entitled to a given number of months of education benefits based on length of time served. With the cost of an hour of flight training being between ten to fifteen dollars per hour, depending on the aircraft used, it is possible a student would lose one week of education entitlement after only one hour of flying. Earning a private pilot license would cause a student to lose over nine months of his education entitlement.[37]

Word of the new rules spread quickly, and enrollment was down significantly by August. The year had begun with 125 flight students but was down to 30 by the end of the summer. Paton met with the university's business manager, comptroller, and veterans' coordinator in an attempt to formulate a plan where students could continue their flight training. This required the university to underwrite the cost of flight instruction since the VA was refusing to pay most flight-fee costs. Prior to the start of fall quarter, it was announced that new financing was in place, and students could expect a normal flight program.[38]

It would take time to see what damage was done to college flight training after the GI Bill rule change and whether students would return even with subsidized fees. Kent State was by no means the only Ohio institution affected by the changes. In January 1948, representatives of five state universities and their flight contractors met in Columbus to formulate an appeal to the VA. They felt that flight courses were being discriminated against, negatively impacting their sustainability.[39] Despite their efforts, it does not appear any relief was given.

Airport Buildings for Sale

Late in 1947, Rudy Van Devere informed President Bowman of his intention to sell his hangar, shop, and office building at the airport. Van Devere was offering the airport structures and property to the university for $75,000, a price he maintained was lower than if placed on the open market. Bowman penned a letter to the State Emergency Board asking for funds to buy the airport facilities, pointing out that among the five universities that bought airports years earlier, Kent State received less funding than others.[40]

The State Emergency Board denied Kent State's request, stating it did not qualify as an emergency. What did not seem like an emergency to those in Columbus felt very much like one to the university. Without owning any facilities at its own airport, Kent State might be forced to deal with an unknown operator if someone else bought Van Devere's property. Kent State was the only university not to purchase buildings as part of its airport acquisition during the war.[41] While that may have seemed frugal at the time, it did not garner any favor when asking for money now.

As this was playing out, Kent State was unofficially approached about establishing an Air Force Reserve Officer Training Corps (AFROTC) unit on campus. This further bolstered Bowman's motivation to secure the buildings at the airport so as not to jeopardize either current operations or future opportunities. Bowman sought release of the needed funds from the State Emergency Board on two different occasions, but he was denied.[42] No further attempt was made to buy buildings that year, nor is there any evidence that Van Devere took other steps to sell his property.

While considering purchase of Van Devere's holdings, the university obtained an appraisal that was completed in March 1948. It detailed the improvements that Van Devere had made over the previous seven years. The buildings included a steel-framed, Quonset-type hangar with aluminum corrugated roofing and an adjoining cement-block hangar covered with an arched wood-frame roof with asphalt roll roofing. Next to this structure was another cement-block and tile building, used as a shop and office. In the rear were a small-frame barn and a storage shed. The appraisers found that the best use of this unique property was for training purposes, suggesting it had a very limited market. It was too far removed from the center of Stow to be attractive for allotment development. Total valuation given for the buildings and two acres of land was $81,680, greater than the sales price of $75,000 that Van Devere had quoted in 1947.[43]

In May 1949, the university received a letter signed by Ray Van Devere, Rudy's brother and vice president of Akron Airways, informing the university that Akron Airways had sold its interest at the KSU Airport to Repair and

Rudy Van Devere's buildings at the airport showing the terminal building and the maintenance and main hangars, 1948 (Kent State University Libraries. Special Collections and Archives)

Repair and Sales Depot, Inc., Ray Van Devere's business, provided flight training, maintenance, and charters, 1956. (Van Devere Family Collection)

Sales Depot, Inc. Ray Van Devere was the president of this new corporation. The university released Akron Airways of its responsibilities and formulated a new agreement with Repair and Sales Depot. In October, President Bowman reported to the board of trustees that he had learned Rudy Van Devere, still the owner of the property, was planning to list his airport parcel with a real estate agent. Although interested in the facilities, the university had no available funds, so no offer would be made. If placing his property for sale had been a bluff on Van Devere's part, nothing came of it, and ownership did not change.[44]

Flight Training Stagnates

It was clear in 1949 that enrollment in flight training at Kent State would not approach the level seen during the immediate postwar years. At that time, veterans were mostly taking flight training as an extracurricular course, not as part of a formal degree. Once the rules changed regarding how veterans could spend their GI Bill funds, only a few remained in the courses. Kent State continued to pride itself on a strong aviation program that offered more than just flying. The aeronautics degree program concentrated on maintenance engineering rather than flight training since trained mechanics were in short supply, while there was an overabundance of trained pilots after the war. Flight training was included in the curriculum to acquaint the aeronautics students with the fundamentals of flying.[45]

Summary

Although the university's experience with the prewar and wartime flight-training programs and ownership of an airport prepared the way for the introduction of Kent State University's own flight curriculum for the average student in 1946, the key issue continued to be that the university did not own the airport facilities. As it continued to work to establish its flight-training curriculum, Kent State's lack of operational control at the airport would repeatedly present itself as an obstacle to those efforts.

Chapter 8

Aviation Takes Hold, 1950–62

In 1948, Kent State University had been approached unofficially about establishing an Air Force Reserve Officer Training Corps (AFROTC) unit on campus. Three years later, when the air force formally announced its plan to create sixty-two collegiate AFROTC units, the board of trustees authorized KSU president George Bowman to apply. Kent State received approval in late April 1951, and its AFROTC unit was activated on July 1, 1951, as Detachment 68. The AFROTC unit would offer three types of training: administration and logistics, general technical, and flight operations.[1] For the first time, Kent State's ownership of an airport and established history of flight training played a role in securing a new university asset.

It was expected that AFROTC flight training would be subsidized, suggesting to airport manager Andrew Paton that it offered the perfect opportunity for Kent State to establish its own flight-training program in lieu of using a contractor. Paton wrote to Dewey Barich, head of Kent State's industrial arts department: "The chief reason for giving our own flight training is the professional one—instruction in any area should be given by educators for education's sake and not by a commercial organization for profit. Neither techniques nor standards are often of university level where personal gain is the sole motive." He also noted that the revenue could go to improving the airport and expanding needed educational facilities. Paton estimated that during the past twelve years of subsidized flight training at Kent State, Akron Airways, Inc., had grossed approximately $312,000. The training had often been unsatisfactory due to turnover and the general low caliber of the contractor's instructors.[2]

When Paton suggested that Kent State offer the AFROTC flight instruction, President Bowman only agreed that flight instructors should be employed by

Kent State to allow for quality control. But he was not as optimistic as Paton about the advantages of having a university-run program of flight training: "I have never shared Professor Paton's optimism over the advantages to the University which might have accrued had we operated our own flight training after our field was useable."[3] Paton's advocacy for a flight program and its use of the Kent State University Airport would continue throughout Bowman's presidency.

AFROTC Flight Training

By spring 1952, flight training was part of the AFROTC cadets' senior year. The university anticipated that by 1954, a minimum of seventy cadets would need flight training, with the number possibly growing to over one hundred annually. Clearly, AFROTC could be a big part of KSU flight training in the future.[4]

The Kent State University Flying Club was also offering flight instruction at the airport, providing an economical alternative to Ray Van Devere's Repair and Sales Depot, Inc., which was the university's contractor at the airport for flight training. The club charged $1.50 per hour for instruction and $3.50 per hour for an airplane, half the commercial cost. This benefited students who needed flight training—aeronautics majors for their private pilot license and non-aviation majors in the elements of flight class. The KSU Flying Club served as a major provider for flight training for Kent State and would continue to do so for years to come.[5]

The eighteen-member KSU Flying Club was one of the oldest, nonsubsidized organizations of its kind in the State of Ohio, and one of the few that actually owned its airplanes. It now had two airplanes: an 85-horsepower Aeronca Super Chief and a 75-horsepower Piper Cruiser. In 1951, during June and July, the KSU Flying Club flew 200 hours for 16,000 miles. To join the club, a student bought a share of stock for seventy-five dollars and then paid monthly dues of three dollars. While knowing how to fly was not a requirement, members did have to take the theory of flight and weather courses. The flight team fielded by the club routinely participated in regional and national events. In 1950, the KSU Flight Team won the national meet held at Stephens College in Columbia, Missouri.[6]

The KSU Flying Club presented pins to its members to reflect their achievements. Each pin consisted of a pair of wings with a "K" in the center on a blue circle background. A bronze set of wings showed a pilot as being presolo, silver meant a successful solo, and gold indicated a pilot had achieved

Flyer, ca. 1927, advertising "Ride in the new Waco Safety Plane," at Stow Aviation Field, using a 1922 Waco stock publicity photo taken at an unknown location (Courtesy of Stow-Munroe Falls Public Library, Fred Long Collection)

Frankie G. Renner's 1930 Waco CRG biplane (NC600Y) used in her altitude attempt as it looked in 2016 (Andrew Heins Collection)

Above: Spirit of the Flying Deuce, a 1968 Cessna 150H, was given to the KSU flight team in 2009 by senior guest flight student Calvin Carstensen to celebrate his wife's recovery from cancer. He chose the name knowing he would never be a flying ace. (Photo by William D. Schloman)

Right: KSU Airport dispatcher handling student airplane assignments, November 1985. *Left to right:* Chris Cornely, unidentified person, Ron Chilleo, Rich Gertz, and Brian Peterson. (Kent State University Libraries. Special Collections and Archives)

Below: Dispatch area inside the KSU Airport terminal, 2017 (Photo by William D. Schloman)

Richard Schwabe with coffee cup by the line service counter in the KSU Airport terminal (Scott and Heidi Koeppl Collection)

Left: Lineup of all available KSU airplanes for Richard Schwabe tribute flight on April 1, 2011. Five Cessna 172s flew in formation followed by the others in line to Cuyahoga Falls, flying over a memorial service for Schwabe. Alumni announced their graduation year over the radio as they passed overhead. (Schwabe Family Collection)

Below: Alumni and friends who participated in the Richard Schwabe tribute flight, April 1, 2011 (Courtesy of the KSU Airport, Beckwith Collection)

Kent State University Cessna Aircraft Liveries

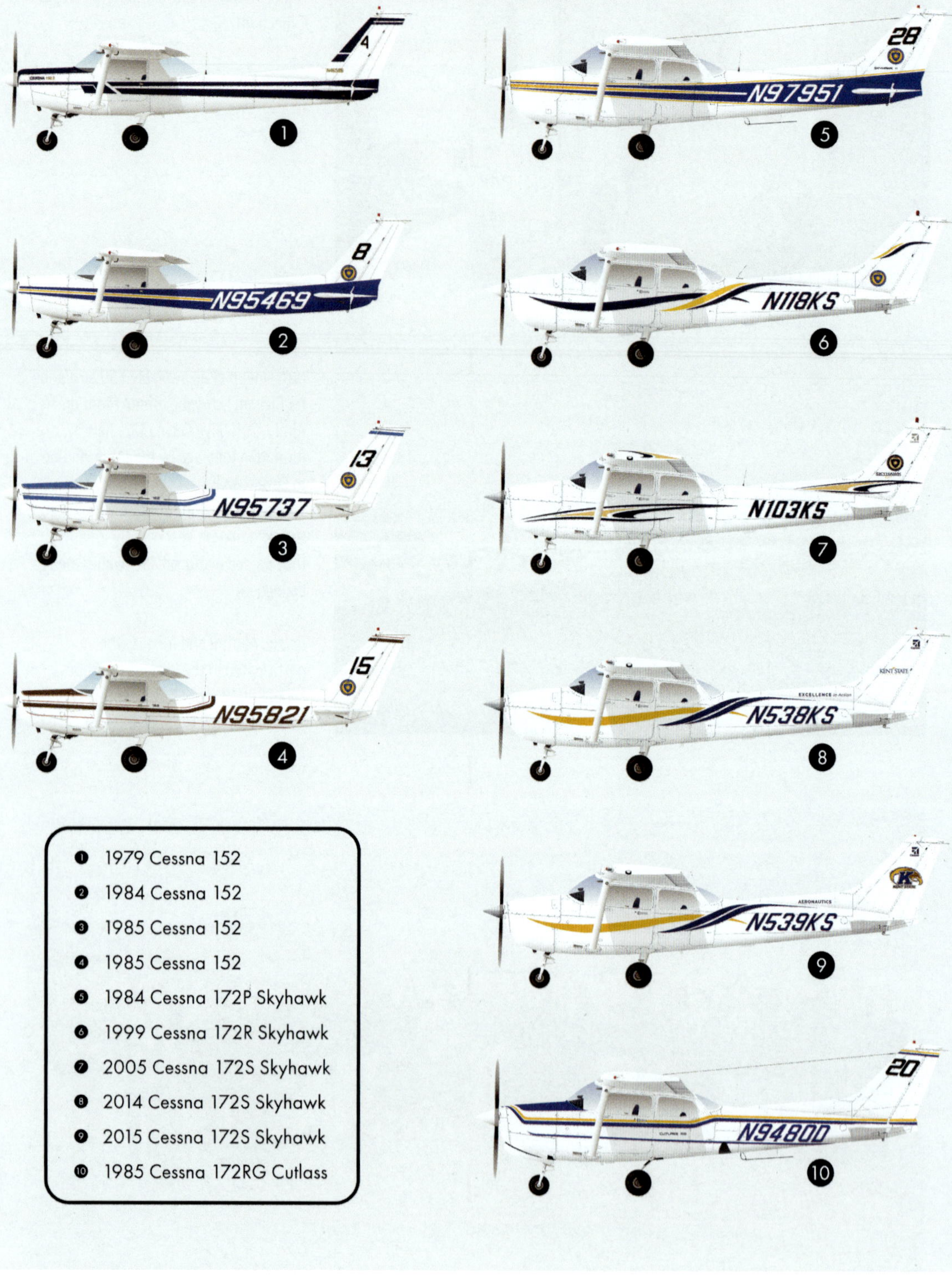

- ❶ 1979 Cessna 152
- ❷ 1984 Cessna 152
- ❸ 1985 Cessna 152
- ❹ 1985 Cessna 152
- ❺ 1984 Cessna 172P Skyhawk
- ❻ 1999 Cessna 172R Skyhawk
- ❼ 2005 Cessna 172S Skyhawk
- ❽ 2014 Cessna 172S Skyhawk
- ❾ 2015 Cessna 172S Skyhawk
- ❿ 1985 Cessna 172RG Cutlass

Drawings by William D. Schloman

a commercial license. A KSU Flying Club tradition required a pilot to wear a parachute to class all day following his solo flight.[7]

Steps toward Final Airport Purchase

In 1952, Rudy Van Devere, who had become a well-established auto dealer in the Akron area, gave notice to Bowman that he was planning to sell his airport parcel with its hangars and was giving the university the first opportunity to purchase.[8] He had also wanted to sell in 1940, 1947, and tentatively in 1949. Previous requests by the university to the State Emergency Board for funding had been denied. Bowman wanted to obtain more input on the airport's value and potential use before making a recommendation to the board of trustees.

To understand the airport's academic potential, Bowman asked E. W. Tishchendorf, head of Kent State's industrial arts department, and Paton to lay out the existing and anticipated use of the airport as part of the department's aviation training program. In their report to Bowman, they indicated that one hundred students were getting experimental flights each year to strengthen their studies in theoretical aeronautics. Pilot training through the department had been given to only six students that year, while the KSU Flying Club was training fourteen members with its airplanes. No course credit was given through the club, but it was seen as "definitely a university function and has distinct educational value." In addition, the students in the Aviation

Move of the Department of Industrial Arts from the heating plant to the new arts building, December 1950 (William Andrew Paton Family Collection)

The arts building, renamed Van Deusen Hall in 1956, early 1950s (William Andrew Paton Family Collection)

Technology Division were now getting practical experience in live aircraft at the airport. The program was still ramping up with the move into the new arts building, but enrollment was expected to reach eighty. Each of the majors would be required to earn a pilot's certificate before graduating.[9]

In his quest for input, President Bowman also asked other state universities about the use of their airports. Ohio State appeared to be the only one making any use of its airfield, while the others characterized the properties as "white elephants" with potential value only for national security and use in time of war. The data gathered showed the investment in Kent State's airport had been only slightly more than half of the lowest reported figure from the other four schools. In addition to funding for land and improvements, each of them had received support for buildings and hangars, which Kent State had not. The final figures showed that Kent State had received $52,000, and the next highest was Bowling Green University with $98,800. Ohio State University was at the top of the list with $450,900 total funds received. The glaring difference was that Kent State was the only one of these universities without its own airport facilities.[10]

In his subsequent report to the board of trustees, Bowman indicated his support for purchase of Van Devere's property. However, he could not support the use of federal rotary funds, as suggested previously by the State Emergency Board. Those were designated for a new university school. Rather, he suggested an appropriation be sought from the state legislature in its next session in keeping with its role in support of national defense. He saw no current or future use of the airport as part of the instructional program of the university. Bowman felt this was an issue that needed to be addressed by the state legislature as a matter of national defense and not internally by the board of trustees.[11]

Rudy Van Devere was informed of the board of trustees' discussion about the property and that the general sentiment was his asking price of $70,000 was unacceptable. Outright appropriations would be sought to fund the purchase. At its next meeting, the board advised the president to obtain a fair market value for the property through "reliable and impartial sources."[12]

An appraisal of the structures and land in February 1953 reported a fair value of $40,000 for Van Devere's property and facilities. One of the appraisers suggested that it might be a better investment to buy Mid-City Airport, which had an asking price of $105,000. However, the final recommendation was that university efforts should focus on the current airport, as it provided greater accessibility for KSU faculty and students. The appraisal also noted that the location of the property limited its use to another buyer without additional land or a needed easement, making such a sale problematic.[13]

With this information in hand, Bowman recommended the board of trustees make an offer of $40,000 for the property. Van Devere did not warmly receive the offer and replied that it "struck me as a typographical error." While disappointed, he closed his response by stating he was willing to consider a future offer.[14]

Bowman reached out to H. Lee White, assistant secretary of the air force, looking for guidance on the necessity of the purchase or even of keeping AFROTC at the KSU Airport. Bowman suggested to White that the AFROTC flight training could be moved to the Akron Airport or the larger Akron-Canton Regional Airport. While White felt he could not give definite advice for what Kent State should do, he did remind Bowman that one of the criteria for establishing AFROTC in the first place was having access to suitable facilities for flight training. This comment may have motivated Bowman and the board of trustees to agree to put forward a new offer of $50,000 for the property.[15]

The second offer was sent to Van Devere on May 5, and on May 15 the university was notified that the offer was accepted. The transfer of the title for the property to Kent State occurred on December 30, 1953. For Paton, this ownership meant "a much richer opportunity than we have had in the past to use the airport as part of the training program instead of just a landing field."[16]

Now the sole owner of all property and buildings at the airport, Kent State's contract with Repair and Sales Depot changed. Ray Van Devere was no longer the owner, but the tenant. Each month he paid rent to Kent State as well as 5 percent of gross revenue from KSU students' flight-training fees. But as the contractor, he had control of airport operations. Even though Kent State now owned both the airport and its buildings, it would find its hands tied to get access to its buildings for its own needs. This new relationship with Van Devere was often contentious in the years ahead. Ownership also meant the university was now responsible for maintenance. Within that first year, the hangar roof showed signs of failure. The airport maintenance fund could not cover a total repair, so the board of trustees authorized a stopgap measure of painting the roof of the hangar and service building with one coat of asphalt-base aluminum paint instead.[17]

Korean War GI Bill

During this same period, Kent State was approved by the US government to provide flight training to Korean War veterans, one of the first locations chosen in the country. The GI Bill would pay 25 percent of a veteran's flight fees, which was an improvement over the revised payment in the World War II GI Bill. In addition, it did not prohibit flight training from being used for avocational or recreational purposes as the World War II program had done. The postwar federal programs greatly impacted flight training at Kent State as well as at the university at large. When the GI Bill expired in 1956, more than 10,000 veterans of World War II and Korea had studied at Kent State. This was more than at any other Ohio institution except for Ohio State.[18]

Partnership with Goodyear

Kent State and Goodyear partnered in 1953 to create a lighter-than-air aircraft training program, making it the only university program of its kind in the country. The program ran from January 19 through July 10, 1953, and was set up specifically for Goodyear to supplement its own in-house training. Andrew Paton was given the role of supervising the program on campus

Andrew Paton and others in a free balloon basket at Wingfoot Lake, 1953 (William Andrew Paton Family Collection)

and assisting specialists in the field. A total of twenty-four blimp pilots were trained during the summer of 1953.[19]

Goodyear also donated an amphibian airplane to the KSU industrial arts department in July 1951. The GA-2B Duck, a three-seat amphibian airplane, arrived on campus in flyable condition but was used purely for classroom instruction. This KSU acquisition was the only Goodyear Duck owned by anyone other than Goodyear.[20]

The KSU aeronautics program saw strong additions to its faculty during this period. Two men would become key contributors to the aviation program for the next two decades. One was Peder Otterson, a University of Minnesota Civilian Pilot Training Program graduate and a flight instructor for Chinese cadets during the war. After release from active duty, Otterson finished his bachelor's degree at the University of Minnesota and then earned a master's degree at the University of Illinois. In 1953, he was optimistic about the future of aviation at Kent State and envisioned an expansion that would include the university owning its own airplanes and aviation mechanic shop at the airport. Otterson was a major in the Air Force Reserves, and in 1968 he earned a doctorate in aerospace studies at Ohio State University and became Kent State's director of aerospace technology.[21]

The second key contributor joining the faculty was Marshall W. Garrett, who had learned to fly at Stow Airport in 1938. Beginning in 1943, he taught at the Akron School of Aeronautics and then served as its principal until he came to Kent State in 1955. He earned his bachelor's and master's degrees at Kent State and would serve the aerospace technology division in several capacities over his career, including as airport manager from 1962 to 1965.[22]

Waco Restoration

In spring 1954, the university purchased a Waco biplane for $150. In dire need of refurbishment, the single-engine, open-cockpit plane provided valuable textbook and practical experience for aviation technology majors, most of whom would be hired as aircraft maintenance engineers upon graduation. Through their classes, the students' work on the Waco included an engine overhaul, structural repair of the wings and body, and re-covering and refinishing the exterior. All this was done under the guidance of Otterson, a certified aircraft power plant mechanic, and Garrett, a certified airframe inspector. At completion, the plane's value was expected to increase to $1,000.[23]

On June 9, 1955, the long-anticipated test flight of the rebuilt Waco took place at the KSU Airport with Otterson at the controls. The plane had been brought from the laboratory in the arts building (renamed Van Deusen Hall

Right: Waco UPF-7 restoration inside the arts building, 1954 (William Andrew Paton Family Collection)

Below left: Waco test flight preparation, Kent State University Airport, June 9, 1955 (Peder H. Otterson Collection)

Below right: Peder Otterson, Waco test flight pilot, June 9, 1955 (Peder H. Otterson Collection)

in 1956) to the airport a month earlier for rigging and assembly. Regrettably, Paton was teaching a class and missed the successful flight.[24]

Glider Training

Ohio Lt. Gov. John W. Brown announced in January 1954 that the state would be opening a glider training school at the KSU Airport. News of the school was not well received by either Rudy Van Devere at Kent State or B. E. "Shorty"

Fulton at Akron Airport. Neither believed gliders could safely operate at airports alongside powered aircraft, nor did the KSU Airport have the hills needed to launch the gliders without towlines. Perhaps there was discussion of a program like the Civilian Pilot Training Program, where students would be given free training, but Van Devere had an issue with that as well. "We've found that students who pay for their flying lessons are much more interested and make better students than the ex-GI's who got their courses free."[25]

Coinciding with that announcement from Columbus, the KSU Flying Club obtained a glider with assistance from the Ohio State Aviation Board. The Schweizer SGU 2–22 glider was purchased for $250 from another flying club. Needing some repair, it was taken to the aeronautics program laboratory on campus for restoration. Despite this attention to gliders, no training school materialized. The only reported glider activity was a flight at the KSU Airport in June 1954, when Lt. David McKay stayed aloft for five hours and ten minutes. At the time, it was believed he set a record for glider flight for the area.[26]

Reconsideration of the KSU Airport's Role

While opportunities and activities for the aerospace technology program were booming during the first half of the decade, Ray Van Devere felt his situation at the KSU Airport was quite different. His simple request on July 1, 1955, would start a nearly decadelong investigation into the future of the airport for Kent State. On that date, he requested a reduction in his rent at the airport from $250 to $200. Business had fallen off, making it difficult for him to afford the current rental rate. In the last year, he had seen his revenues drop 50 percent and wanted assistance from Kent State to maintain the field. While Paton agreed with his claim that business was slow, Bowman had his doubts that it affected his ability to make a profit. As he saw it, Van Devere was not paying property tax and the airport was one of the busiest in northeast Ohio. Rather than lower his rent, initial consideration was given to bidding for a new contractor.[27]

Paton, who had been dealing with Ray Van Devere since 1941, didn't support a rental reduction either. Van Devere was making money from fuel sales and airplane storage, sales, and rentals. Paton saw this as a moment to reevaluate the future of the airport and its educational use. As it stood, they didn't even have access to the buildings they owned, as they were occupied by Van Devere's Repair and Sales Depot.[28]

Paton saw three options for the future of the airport. The first was that the university could proceed as it was, keeping classes on campus and limiting activities at the airport. The second option called for the university to establish educational facilities at the airport to provide flight instruction, along with a commercial operator to serve the public. The third and most expansive option was for the university to take over the airport's entire operation and close it to commercial and public operations; all facilities would be used solely for university training purposes.[29]

For the airport to be an educational facility, even under the current arrangement, the university had to figure out how students could get greater benefit from it. With grass runways, the airport was nearly unusable from February to May. If it were possible to secure federal funds to install a 3,500-foot hard-surface runway, the airfield would be safe for flying for more of the year, and revenue would increase. At the time, most of the eighty aviation majors were obtaining much of their required flight training elsewhere. The board of trustees concluded that rather than reduce Van Devere's rent, the airport should be improved. In the short term, cinders could be acquired from Goodyear to improve runway and taxiway conditions during poor weather. In the long term, the biggest issue would be to change Ohio law so universities could access federal funds.[30]

Paton's suggestion was to cancel the university's contract with Repair and Sales Depot and find a new contractor. He felt Van Devere was not suited to operate an airport, nor was he interested in it as an educational institution. Fees for his flight training were the highest in the region. Paton's argument mentioned the experience of the Eagle Squadron, an army and air force flying club formed at Kent State in 1954. Its goal was to give cadets added preparation for the military and to offer cheaper flying rates, but they had not been welcomed by Van Devere to locate at KSU Airport. Receiving a much different reception from the manager of Mid-City Airport, the club chose to hangar its airplane and take instruction there. Later, Van Devere asked that the club reconsider KSU Airport, but because the rates would still be higher, the club elected to stay at Mid-City.[31]

Mid-City Airport Closure Impacts Kent State University

The aviation landscape for the region changed in January 1956 with notice that General Motors (GM) had purchased Mid-City Airport. The airport would be torn down to make way for a new factory for GM's Euclid Division. This closure marked the end of a facility that had played a key role in area aviation since 1929, including the training of hundreds of KSU students since 1939. Operations at Mid-City ceased on March 30, 1956, displacing four hundred pilots who used the airport. At the time, Mid-City was providing most of the flight training for KSU students, placing additional pressure on Kent State to sort out its airport dilemma. The closing forced the Eagle Squadron to move its Piper Cub to the KSU Airport.[32]

Following GM's decision, Paton found himself back in Columbus in March 1956 to meet with the director of finance for Ohio and an aviation director and engineer from the Civil Aeronautics Authority (CAA). The primary discussion involved getting a hard-surfaced runway suitable for industrial use, and possibly for the new GM facility, and how it could be paid for. The goal was for a paved 3,000–3,500-foot runway that could be used by twin-engine aircraft similar in size to a DC-3 airliner. The price quoted put the total project at $150,000. The important question was whether Kent State could get access to federal funds. The answer was still no. The opinion of Ohio attorney general C. William O'Neill was that the board of trustees was running the airport for the university and not for public needs. The university's lack of authority to maintain an airport for the public prohibited receiving CAA funding. This funding obstacle remained an issue for Kent State, and it would take years before a solution was found.[33]

With the start of winter quarter 1957, Paton reached out to Bowman for resolution of the airport dilemma. The university now owned the airport buildings but lacked access because they were still leased to Ray Van Devere and his Repair and Sales Depot. Taking inspiration from Ohio State, Paton preferred taking over both the buildings and operations from the contractor. This would give Kent State full freedom to use the airport and its facilities.

Groundbreaking for General Motors' Terex factory at the site of the Mid-City Airport, March 27, 1957. Gov. C. William O'Neill is in the earthmover. (Hudson Public Library and Historical Society)

Above left: Stow Airport Flyers played on a baseball field west of the airport hangars known as Stow Field, 1951. Billy VanGunten (*third row, third from right*) graduated in aerospace technology from Kent State University in 1964. (William VonGunten Collection)

Above right: Newspaper ad for Van Devere Aviation, formerly Repair and Sales Depot, *Akron Beacon Journal,* April 20, 1962 (Courtesy of the *Akron Beacon Journal*)

Right: House across from the KSU airport on Kent Road destroyed by crash of Ray Van Devere's airplane, 1959. (Courtesy of the *Akron Beacon Journal*)

Ideally, the airport would be made private, but that would not be advantageous to the aviation community. If Kent State kept its contractor, the university should then build its own airport facilities. In any case, the future of a paved runway necessitated that the university buy land between the current property and the east-west road to the south. Bowman felt the funds were available to buy the necessary land, even though there was no expectation of a new runway any time soon. And he again expressed that he was not interested in taking over airport operations.[34]

Airport Improvements

By 1960, the Kent Chamber of Commerce was interested in improving the airport to encourage local industry to use it as a base for their executive airplanes. Several area businesses had based their airplanes at other airports because they could not use the KSU Airport due to a soft field or a lack of lights at night. Chrysler, for example, surveyed the airport but decided to base their airplanes in Cleveland. Meanwhile, KSU officials were reluctant to make the airport a public field, thus restricting them from getting federal funding.[35] It had been five years since the first discussion of improving the airport, but no headway had been made.

The Federal Aviation Act of 1958 created the Federal Aviation Agency, replacing the CAA. This new agency would be renamed again in 1967 as the Federal Aviation Administration (FAA). The 1958 law changed the requirements to earn a private pilot license, adding several hours of instrument training. This impacted the cost of flight training for a pilot license as now the university needed an airplane with appropriate instrumentation. Until the university adjusted the class fee, the GI Bill wouldn't cover the increase.[36]

While improvements for the airport were stalled in October 1960, the board of trustees requested an appropriation of $360,000 from the state for a new aviation education building at the airport. If successful, the entire aviation technology program would move to the airport. This offered great promise for KSU's aviation technical program, which already had eighty student majors, each required to take flight instruction. A new building at the airport would serve the current needs, as well as offer new opportunities, and would serve as a "shot in the arm" for the program.[37]

In spring 1961, Paton sent a proposal recommending a change in airport operation to John J. Kamerick, dean of the College of Fine and Professional Arts. Given the poor experience with both Rudy and Ray Van Devere in operating the airport and the barriers created to its educational use, he recommended again that the contract with Repair and Sales Depot be canceled and that university staff take over airport operation beginning with fall 1961. He reiterated the now-familiar issues and laid out an implementation plan.[38] This proposal would give Kent State its own flight program, but it still needed acceptance within the administration.

An immediate problem going into the summer of 1961 was that the hangar was in serious need of repair, having lost all of its protective coating on the steel frame. Approval was given to apply a new prime and finish coat over the interior steel. There was also news that budget action by the state legislature killed any chance of Kent State getting financing for the airport. As

KSU vice president Robert I. White put it, "legislative budget action is always conclusive and, like death, so final."[39]

The following April, the university bought an army surplus, twin-engine Beechcraft C-45. It was flown in from Tucson, Arizona, by Peder Otterson and temporarily stored at Akron Airport because the grass runways at the KSU Airport were covered in mud, making landing impossible. The size of the airplane made it too expensive for student flight training, so plans were to use it for ground instruction.[40]

In spring 1962, approval was given to change the title of the Aviation Technology Division of the Department of Industrial Arts to the Aerospace Technology Division. This was the last action affecting the program during Bowman's tenure as president.[41]

Summary

By 1962, Kent State University at last owned the airfield, the hangars, and the terminal for their airport. To a large degree, this did not change how the airport could be used. The contractor still controlled the operation and could limit access, causing mounting frustration. Ownership brought increased responsibility for maintenance and capital improvements. The university explored federal funding but remained hesitant to take funds that would require the airport be operated for public, as well as campus needs.

Despite lack of full airport access, there were memorable moments for KSU aviation. The university was awarded an AFROTC unit. Hundreds of Korean War veterans were trained under the GI Bill. With the expertise and direction of Peder Otterson and Marshall Garrett, students rebuilt a Waco biplane and a Schweizer glider. Relations with Goodyear led to a lighter-than-air training course and gift of an amphibian airplane. A sad note came with the closing of Mid-City Airport, an important part of KSU flying for fifteen years.

During this period of total KSU ownership of the airport, Andrew Paton was a consistent voice for the university to take over airport operations and provide its own flight training. Despite not sharing Paton's vision, KSU president George Bowman valued his counsel: "You are one of the stalwarts in the organization upon whom I know I can always count for competent and thoroughly thought-out advice."[42] Ray Van Devere's request for a rent reduction spawned a reevaluation of the airport and its purpose that over the next years would eventually lead to change.

Chapter 9

Total University Control at Last, 1963–71

Leadership at Kent State changed in 1963 with Robert I. White becoming university president. He inherited an airport that still had a largely undefined role. Andrew W. Paton and the Aerospace Technology Division faculty were strongly interested in the airport becoming an educational facility and for Kent State to take over flight training from its contractor, Van Devere Aviation (formerly Repair and Sales Depot, Inc.). Frustration with contractor Ray Van Devere had been building over restricted access to the facilities and the cost and quality of flight training offered. In addition, the airport suffered from deferred maintenance and an overdue investment in capital improvements.

J. W. Bunn, vice president for business and finance at Kent State, shared with White that trustee Robert Dix, a pilot, had been pushing for runway improvements for ten years. Bunn believed that KSU president George Bowman's restraining influence had stopped any airport additions or improvements. Bunn also had his doubts about airport development and expressed his willingness to investigate both what was needed and what funds could be had.[1]

Bunn reported his preliminary findings two months later, concluding that the existing airport was inadequate. The runways were below acceptable standards for safety and all-weather use. Additional property would need to be purchased, and encroaching real estate developments would make expansion difficult. Federal Aviation Administration (FAA) funding would pay one-half of the costs but could not be used for hangars. Kent State could apply at any time, but funds would not be available until July 1965. Bunn suggested consideration be given to purchasing a new site.[2]

The Airport Special Committee

White made a decision that would once and for all establish a future for aviation at Kent State. In October 1963, he formed a University Airport Special Committee to produce a thorough analysis to determine the long-term future for the Kent State University Airport and "to examine the possibility of moving to a higher range." The committee was chaired by Robert E. Hill (dean, business administration) and included Bunn, Emil Berg (business manager), Elmer Novotny (interim dean, College of Fine and Professional Arts), Lt. Col. Manley Morrison (commander, AFROTC), and Andrew Paton (airport manager and director, Aerospace Technology Division).[3]

The committee was charged to look at possible expansion of the airport, investigate what improvements would make the airport more valuable and serviceable, and identify what greater service the airport could perform for the community and regional users. Immediate plans to pave the runway were shelved as they were deemed a "temporizing, too-little project" in the face of possible major improvements. White believed that "for years we have been operating a port which has been surprisingly useful in general aviation but which, nevertheless, falls far short of a large potential service to the University's educational operations, to University general service, and to the functions of a really useful regional airport. We have only done that which is enough for minimum purposes and, I hasten to add, for good reason in the face of the financial situation." The committee was given a deadline of winter quarter 1964 to produce a final report.[4]

Prior to formation of the Airport Special Committee, White asked Paton for a brief history of the KSU Airport, which he then shared with the board of trustees in announcing the formation of the committee. Before taking medical leave in December, Paton authored a second report on the pattern of airport use. These two documents would provide the Airport Special Committee a starting point for its work.[5]

Paton's reports outlined that the KSU Airport in late 1963 was classified a Class II sod field with two runways, 3,000 feet and 2,400 feet long. Nearly all flight training at the airport was provided by the Kent State University Flying Club and Van Devere Aviation. Kent State owned two airplanes, a Stinson used for demonstration and introductory flights and a nonflying military surplus Beechcraft C-45 used for practical experience.

While the university now owned the airport and all the buildings, it was still limited in using the airport as it saw fit. The contract with Van Devere Aviation provided student flight training and gave Ray Van Devere the use of the hangar and office building for his purposes. Any time the university

Professor Marshall Garrett's class, Van Deusen Hall, 1964 (Kent State University Airport)

or KSU Flying Club wanted to work on one of their own airplanes, it would have to be disassembled, driven to campus, and reassembled in the department's laboratory.[6]

Dissatisfaction with flight training was also an issue. The Air Force Reserve Officer Training Corps (AFROTC) program started its flight-training program with Van Devere during the 1959–60 school year but found the training so unsatisfactory that it canceled its agreement before completion. At the time the committee was beginning its work, twelve Reserve Officers' Training Corps (ROTC) students were learning to fly at the airport and expected to be the first successful group to finish their licenses. Meanwhile, twelve KSU students had chosen to fly through the KSU Flying Club since the cost was one-half that of Van Devere Aviation.[7]

With a deadline only a few months away, the committee began work immediately and updated the president and board of trustees every few weeks. Previously discussed short-term plans, like blacktopping the runways, were scrapped to ensure that any money spent provided a long-term return. Initial comments from the FAA showed the KSU Airport needed an estimated $300,000 worth of improvements to meet its safety standards. During the previous decade, money had not been available from the government, but starting in 1965 the FAA would pay for 50 percent of such improvements. The FAA would also reimburse Kent State for capital improvements used for educational purposes going back as far as 1946. A key stipulation of accepting FAA funding was that the university would be required to commit the airport to a minimum of twenty years of use.[8]

The runway, both its surface and length, was the focus for airport improvements. The educational use of the airport was hindered by the grass runways, which in northeast Ohio were unusable at least two months a year due to winter conditions. This greatly affected reliable flight-training conditions and service to the general aviation community. Presently, the longest runway was 3,000 feet, while the minimum FAA requirement for landing a twin-engine aircraft was 3,400 feet. In addition, an obstruction-free zone of at least 1,200 feet was required at each end of the landing strip. The existing runway was oriented north-south along the length of the airport property. An east-west road ran along both the north and south sides of the airport property. While the university had frontage on the road to the north, it did not own the 34.7 acres to the south, which blocked its ability to extend the runway any further. Any plans involving a runway expansion would first require acquiring land, if available, and possibly an easement to the south.[9]

Death of Andrew W. Paton

The single largest blow to aviation at Kent State came on February 20, 1964, with the death of Andrew W. Paton at the age of fifty-five. Paton had been involved with aviation at the university since its earliest days and served as director of the aerospace technology program and as airport manager. Over the years, he was the most vocal proponent of the airport's value and the key visionary for what Kent State's program should become. In March 1963, he had undergone brain surgery and later returned to his work at Kent State. In December, his health deteriorated, requiring him to take leave. Paton was eulogized as having "success in the field of education, [and] he foresaw the immense potential of aviation ahead of many of his colleagues and he influenced the lives of many young men and women both here and abroad through an unselfish desire to impart his knowledge."[10] The loss of his leadership would be felt in the years ahead.

Marshall Garrett was designated airport manager to replace Paton and also assumed Paton's place as the ex officio member of the Airport Special Committee. The loss of Paton slowed the committee's efforts. Committee Chair Hill stated the overarching goal: "We want to have an academic accredited aviation program there. But we also foresee the evolution of the Airport to accommodate business and industry in the area." Questionnaires were sent out to 1,100 local businesses in seven neighboring towns to gauge the number that used the airport for business. Roughly one-third of the 144 responding businesses said they used it.[11]

Garrett surveyed airlines to determine the industry's need for trained pilots. What he learned was that both commercial and military aviation would have a pilot shortage of 100,000 by 1970, with the air force projecting a greatly increased demand. The flight-training needs of AFROTC at Kent State, however, were dependent on cadet enrollment and the success of recruiting.[12]

Delmar Olson, chairman of the Department of Industrial Arts, sent out inquiries to six of the biggest airlines in the nation looking for their feedback. He shared with them that KSU graduates now found their way into positions at airlines in engineering, research, and management with only a private pilot license. If the university extended the current curriculum, it would be possible for graduates to meet the qualifications required for being an airline pilot while earning a bachelor of science degree. Demand was high for properly licensed pilots in the industry, and Kent State saw this need as one they could help meet. Olson's inquiries revealed that the KSU aerospace program was practically unknown, even on campus. The efforts of the Airport Special Committee could put KSU aeronautics on the map and give students a program that would prepare them for an airline career.[13]

The Airport Special Committee determined there were three possible funding sources to pay for recommended improvements: federal, state, and university. The advantage of federal funding was that 50 percent of the costs would be covered for improvements that met FAA standards, which ultimately was the goal anyway. This option, however, required that the university purchase land to meet FAA requirements. The option to seek state funds seemed problematic due to political issues with state administrators who did not see airport investment as an educational expense. Funding by the university would require using limited local resources, but it would have the advantage of averting federal restrictions, state politics, and likely delays. The committee's recommendation would be for university funding.[14]

On June 24, 1964, eight months after its formation, the Airport Special Committee submitted its final report to President White. The committee made four recommendations that addressed the well-known issues both with the airport and its related educational activities.[15]

The first recommendation was to bring both instructional and operational control of the KSU Airport entirely under university control. This would provide the current aerospace technology program full access to services and facilities at the airport. No longer would a contractor have a monopoly on the university-owned facilities. Most important, the committee recommended the creation of a KSU flight program and elimination of the dependency on a commercial flight operation to meet the needs of their curriculum and students.[16]

To support these changes, the committee's second recommendation addressed airport facilities. Runways, hangars, and facilities should be improved to ensure a safe and all-weather operation as soon as possible. The much-discussed paved runway would allow for twelve months of airport operations and be safer than the grass runways. In addition to improving the current buildings, new facilities would provide for future expansion of the aerospace program. Educational activities would then be focused at the airport. To pay for these changes, the third recommendation called for the needed funds to be allocated by the board of trustees. The final recommendation was to honor Andrew Paton by renaming the airport "Andrew W. Paton Airport—Kent State University."[17]

Airport Named for Paton

On July 9, 1964, the board of trustees, with White's endorsement, approved both the report's findings and the new name for the KSU Airport. The following year, it came to White's attention that the university had neither asked for permission nor officially informed Mrs. Paton of the naming of the airport after her late husband. In spite of this etiquette misstep, Maurine Paton and her family were very pleased by the trustees' unexpected action.[18]

The recommendation that required immediate attention was the improvement of the runway. Before lengthening the runway could commence, it would be necessary to acquire the 32.7 acres of land to the south of Paton Field. The board of trustees approved the purchase of the Belcher property, which was completed two months later for $101,400.[19]

Not everyone welcomed news of the proposed changes outlined in the report. A group of Stow and Munroe Falls residents proposed to the Akron City Council's Airport and Aviation Committee that the university's operations be moved to the Akron Airport. As they saw it, such a move would

Peder Otterson with Andrew W. Paton Airport sign, 1974 (Kent State University Libraries. Special Collections and Archives)

provide needed tenants and revenue to the Akron Airport and save taxpayers money by combining the two operations. The group asked the Akron City Council committee to approach Kent State University with their proposal. Similar arguments were made to Stow City Council, along with concerns that flights by larger aircraft would create a nuisance for their properties. The university held a public meeting in October 1964, giving the concerned residents a chance to air their concerns. Some had the misconception that an east-west runway was to be built. Kent State shared that the projects at the airport were being financed from fees and other university income, not with state funds. Some residents indicated they would continue their fight against airport expansion.[20]

D. E. Dominic, superintendent of Stow City Schools, reached out to Kent State following the announcement of the committee's report and its purchase of the land adjacent to the airport. It is unlikely that the university was aware at the time of their land purchase that the Stow Board of Education had plans for a stadium in the area and that it now saw those plans in jeopardy. An easement would be needed to provide ingress and egress to the stadium. Unfortunately, the purchase of the Belcher property gave lifetime usage rights to the former owners. White reassured Dominic that the university would find some means to accommodate the need of Stow City Schools.[21]

In July 1965, the board of trustees approved $145,000 to pave the 3,000-foot runway and add runway lighting. It is unclear why the decision was made to stay with the 3,000-foot length when it was known that the FAA requirement was for 3,400 feet. Perhaps, even with the purchase of additional land to the south, it was not possible to meet the obstacle-clearance requirements.

Runway lighting was another long-hoped-for improvement. The practice in 1965 was for airport staff to light a smudge pot and set it out to provide nighttime guidance.[22]

Addressing Flight Program Needs

Another issue for a new flight program was the lack of proper training airplanes. The department's Stinson, while suitable for introductory flights, was not going to meet the needs for a proper flight-training program. As a result, plans were made to trade in the Stinson to buy a new four-seat, single-engine airplane. Because the Stinson had also been used periodically for executive flight transportation for university administrators and faculty, the plan was to refurbish an old Beechcraft C-45 twin-engine airplane as a replacement. When Kent State purchased the military surplus C-45 in 1960, it agreed not to sell it for ten years. Originally, it was planned to be a maintenance training tool for aerospace technology majors, but mostly it had sat dormant at the airport, earning it the nickname "Kiwi Bird."[23]

The goal to establish a flight program started with tackling airport improvements, yet attention to building the program itself was initially lax. By May 1965, a director for the Aerospace Technology Division had not yet been appointed, seriously delaying curriculum revisions. The aerospace staff assumed that Peder Otterson would be named. Although he was recommended for the position by the chair of his department as early as fall 1964,

Professor Peder Otterson at his desk, Van Deusen Hall, 1968 (Kent State University Airport)

the position remained vacant. Under pressure to earn a PhD, Otterson reluctantly requested a one-year leave of absence to pursue the degree at Ohio State, although he felt his time would be better spent rebuilding the curriculum. For his doctoral dissertation, he would undertake an evaluation of Kent State's aerospace technology program, identifying industry trends and the criteria of an effective program. There was also general frustration about the lack of progress toward a new education building at Paton Field, despite the Airport Special Committee's recommendation and approval by the board of trustees. Tensions with Van Devere Aviation continued over barriers to using the airport for educational purposes.[24]

Master Plan Developed for Flight

Because little action had resulted to date from the recommendations of the Airport Special Committee, Garrett and Lieutenant Colonel Morrison prepared a master plan in September 1965 to guide implementation. Phase I included all the short-term goals needed to establish a flight program and take over airport operations. The first step was to give Van Devere Aviation reasonable time to liquidate its assets and vacate the airport property. This would allow all flight training to be the responsibility of the Aerospace Technology Division. This new program and all other airport operations would fall under the responsibility of an airport manager. A new terminal building would need to be built on the north side of the hangar for flight planning, a weather station, and flight operation offices. The last steps in Phase I included installing fuel pits and building T-hangars (individually partitioned "T-shaped" hangars for single aircraft) for non-university airplanes.[25]

Phase II consisted of building a new education facility at Paton Field to accommodate the Aerospace Technology Division. The planned building, to be located at the southeast corner of the airfield, would be a 180 x 250 ft. facility with an expected cost of $450,000, to be financed by the university as an academic building. Placing the building at the opposite corner of the airfield from the present facilities would establish a clear separation between university and public use of the airport. Phase III provided for a new maintenance hangar and an expanded parking lot. Phase IV would expand the runway to 5,000 feet long and 100 feet wide.[26]

Two weeks later, the board of trustees approved the long-range development laid out in the master plan. Runway paving was expected to be completed by November 1, 1965, establishing the airport as a year-round facility, which would also allow for FAA certification.[27]

Airport aerial (1972) showing T-hangars, paved runway, and fuel pit, which were completed in 1966 (Kent State University Libraries. Special Collections and Archives)

During fall 1965, the FAA approved the establishment of Part 61 Flight Training courses at Kent State University, giving it authorization to conduct its own flight training. No longer would the university have to rely on a contractor. But they still needed two instructional aircraft and an FAA-licensed flight instructor and examiner. The recommendation for the latter was Leonard Mack, who could also serve as executive pilot for university administrators.[28]

Preparations Made for Flight-Training Takeover

On February 1, 1966, KSU vice presidents John J. Kamerick and J. W. Bunn approved Olson's proposal that the Department of Industrial Arts take over the total operation and management of Paton Field as an auxiliary university enterprise with four divisions of responsibility: flight operations, a general university flight program for students not in the four-year program, flight instruction through the GI Bill and for ROTC, and executive flight service for university administration and faculty.[29]

Yet by April, no formal decision on the flight program had been made. Morale among aerospace students was becoming a serious concern. Their uncertainty as to whether they would be able to get their flight instruction in the fall could lead to their mass exodus if reassurance was not forthcoming. The aerospace majors knew that facilities were inadequate, equipment was obsolete, and some instruction dated.[30]

No airplanes had been purchased, although the recommendation for two training aircraft had been made two years earlier. The KSU Flying Club was still providing most of the flight training for aerospace students, although it had no formal or legal affiliation with the department and operated outside its control. Also, the amount of flying the students needed was greater than the club could provide. A backlog of over 1,100 hours of required flight for students in the freshman sequence existed in April 1966, too much to be completed by the end of term.[31]

Enrollment had recently increased due to promotional work by the university the year before, bringing one hundred students into the aerospace technology program. This put increased pressure on already inadequate resources for both classwork and flying. Friction with Van Devere Aviation meant the university was not able to gain access to facilities, and tensions were building. The department offered that if the necessary operational decisions were made, it would be willing to offer Ray Van Devere a position in the organization.[32]

Ray Van Devere, assistant airport manager, at his desk, 1967 (Kent State University Libraries. Special Collections and Archives)

Emil Berg, business manager, met with Van Devere on July 22 and gave him the official news that he had to relinquish his business at the airport and liquidate his assets within thirty days. Van Devere was offered a position as assistant airport operator. After nearly thirty years at the airport that started with Stow Field in the late 1930s, it was hard to accept this change and become a university employee. He asked for a week to consider the university's offer and then accepted. In other staffing news, Olson reported that a second flight instructor had been hired for the beginning of fall quarter.[33] This was Richard Schwabe, returning to the program where he had learned to fly.

With the start of the 1966 fall quarter only a week away, Scott Henninger, KSU Flying Club

Irma Hanson, airport administrative specialist, recalled sewing wing covers on old airplanes in earlier years, 1968 (Kent State University Libraries. Special Collections and Archives)

president, was notified that the Aerospace Technology Division at Kent State was taking over flight training at the airport. The university acknowledged their appreciation for the long history of the KSU Flying Club providing flight training for students since 1947. While a loss for the club, this new development fulfilled a measure of Paton's vision. The KSU Flying Club was encouraged to continue to operate at Paton Field. This letter to the club included the first mention of the aerospace division owning its own airplanes—most likely purchased from Ray Van Devere.[34]

Olson wrote Kamerick just days prior to the start of fall term about continuing issues. They still had to hire someone to be responsible for the airport; Olson recommended Marshall Garrett, who was also currently in charge of flight operations. Although Mack would continue as executive pilot, they also wanted him to continue to serve as FAA examiner, and the university needed to hire a second flight instructor to help cover the increased flying needs of AFROTC, which would dominate the flight instruction in fall term. Neither

Richard "Dick" F. Schwabe (Schwabe Family Collection)

Richard "Dick" F. Schwabe (1921–2011) grew up in Cleveland and enrolled at Kent State University. Accepted into the Civilian Pilot Training Program (CPTP) in 1941, Schwabe soloed in April and completed the secondary course that fall. He enlisted in the army air corps and by 1943 was flying "The Hump" over the Himalayas between India and China. Allied aircrews flying that route faced constant threat from Japanese fighters and adverse weather conditions, causing one-third to perish. Schwabe received the Air Medal with one oak leaf cluster and the Distinguished Flying Cross. After the war, he returned to Kent State to finish his degree and in 1949 started both a master's degree program and his career as a science teacher and coach in local high schools. In 1966, Kent State hired Schwabe to be the lead flight instructor for its new flight-training program set to begin in the fall with equipment and personnel in short supply. Schwabe expected the best from his students and appreciated their efforts, proudly referring to the men and women in his program as "one of my boys." He was well known for his air show performances and within the aviation community as an FAA examiner. Schwabe's passion for teaching shaped thousands of pilots, and many reached out to him in his last days. Jean, his wife of sixty-five years, read him each card and letter. One struck a chord with him, and he repeated its closing two words as his last: "Mission accomplished." He died on March 23, 2011, leaving a legacy of high standards and excellence.[35]

flight training for the eighty aerospace technology majors nor orientation flights for another one hundred students would be possible. Olson brought up additional issues: the airport terminal building had to be remodeled for use by instructors, students, and flight operations; space at Van Deusen Hall was at a premium with no room for new test equipment soon to arrive; and flight-instruction fees had yet to be set. There was a lot to be sorted out.[36]

Start of KSU Flight Program

Wednesday, September 28, 1966, marked the beginning of fall quarter with Kent State University officially offering the flight-instruction classes.[37] Twenty years earlier, after introducing flight training into the aerospace technology curriculum, Paton envisioned a day when Kent State would have its own program and make its airport an educational facility. While the facilities were still lacking, the heart of the program was in place. Schwabe took the first students airborne that day. He always credited Paton as the reason the flight program existed, so it was entirely fitting that the first flight took off from Paton Field.

On October 20, 1966, President White addressed the board of trustees and announced that Kent State University had taken over the pilot-instruction program. The board approved a rate of $11.88 per hour for flight fees, which compared to $16.50 per hour at other local flight schools. As Paton had predicted, a benefit of the university operating its own flight courses would be less expensive training for students.[38]

A new operator showed up at Paton Field in 1966. There had been some backlash with the removal of Van Devere Aviation and the resulting loss of services to local pilots. The university gave permission to a private pilot, Albert C. Beckwith, to start a flying club at the airport. With a $15,000 investment, he bought a Cessna 172 and started Commercial Aviation Corporation. This began a long relationship with the university. Beckwith would be one of the airport's biggest supporters over the decades to come.[39]

The only fatal accident at Paton Field occurred less than a month into fall quarter. Rev. Dewey Hale, a former KSU aerospace technology instructor, and a passenger departed the airport in a Piper Tri-Pacer. The last contact with the small airplane came when it was flying over the airport and asked for a radio check. Four hours later, Schwabe took off to the south with a student, and while glancing out the right window, he noticed what appeared to be a pile of debris. It was not uncommon for someone to dump junk at the edge of the airport property, but Schwabe radioed airport operations to

Albert C. Beckwith (Photo by William D. Schloman)

Albert C. Beckwith (1927–2014), a New Jersey native, relocated to Stow, Ohio, in 1964 to work for AMOCO Chemicals Corporation. After earning his pilot license, flying became his passion. Beckwith soon purchased his first airplanes, which flew out of the KSU airport, and began his flight business with a flying club to repay his airplane loans. He formed Commercial Aviation Corporation (CAC), which was his first role as a fixed-base operator (FBO), at the KSU Airport in 1966. As his fleet grew, he opened a second FBO at Akron Airport. The fuel shortage in 1976 prompted him to liquidate most of his airplanes and sell his maintenance operation at Akron Airport. In 1977, Beckwith leased land at the KSU Airport for his CAC office. Over the next years, he made several proposals to the university for airport improvements that would benefit both his business's and the airport's financial viability. Beckwith worked diligently to organize a defense for the airport when two KSU administrations threatened its survival. In addition to flying for business and pleasure, he partnered with a friend to build a Bede BD-5B, a small single-seat aircraft, from a kit. Completed after six years, the airplane was flown for about twenty hours and then donated to the Smithsonian, where it is now displayed. Beckwith is remembered for encouraging others to think "outside the box" to achieve their greatest potential. He was a mentor to many flight students and an inveterate promoter of aviation at Kent State. Beckwith died on October 1, 2014.[40]

investigate. Upon arriving, the airport personnel found the downed airplane, which appeared to have landed short of the runway and flipped over, killing both occupants.[41]

As the quarter continued, the university's strained relationship with Ray Van Devere reappeared. Van Devere had accepted a position as an assistant airport operator, but the permanent position of airport manager was still vacant. Van Devere, unhappy to have lost his flight-training contract, made a habit of openly criticizing the university's program. He was known by some as a "grass roots flyer" and not taken with the protocols of the "new flyer types." Frustration with this situation spurred interest on campus to permanently fill the vacant manager's position and hope Van Devere might choose to quit. It would take until January 1967 for the position of airport manager to be filled by George Hoagland, former mayor of Tallmadge.[42]

During this time, the position of director of the aerospace technology program was finally filled with the appointment of Peder Otterson. Until his retirement in the summer of 1979, Otterson was a strong voice for improvements to the program. He became chair when the Aerospace Technology

George Hoagland, airport manager, and Richard Schwabe, chief flight instructor, 1967 (Kent State University Libraries. Special Collections and Archives)

Division became an official department and would later serve as assistant director for the School of Technology.

Several new additions came with the winter quarter of 1967. The University Council and Curriculum Committee approved four new courses: Commercial Flight phase I, II, and III and instrument flight. Shortly thereafter, Kent State took delivery of two new airplanes, a Cessna 150 and larger Cessna 172, which would be needed for the new instrument flight course. James "Smitty" Smith was hired for the winter quarter as a flight instructor to help Schwabe with the growing student load.[43]

With the flight program finally established, attention returned to building a new educational building at the airport. The planned building would separate university flight operations and general use operations, requiring that the facilities include a hangar for university airplanes, laboratories, classrooms, and maintenance facilities. These would serve both flight-training and aerospace engineering technology programs. The proposed building would be 70,000 square feet, enough to handle 800 to 1,000 students, and include a 150-square-foot hangar. The building configuration was similar to that in place at air force technical schools around the country.[44]

Flight Program Continues to Grow

By summer 1967, two part-time flight instructors were added to the roster—Eugene G. Ripple and Richard Polinski. Ripple had been a US Marine pilot for sixteen years, serving in both the Korean and Vietnam Wars. He would

Eugene Ripple, flight instructor and later airport manager, at his desk, 1969 (Kent State University Airport)

be a member of the aerospace faculty for the next twenty-five years, including twenty as airport manager. Because of budgetary problems, the university considered shortening the flight instructors' contracts to nine months, omitting the winter quarter when weather was at its worst. Fortunately for the flight instructors, their current contracts were through 1968 and could not be adjusted. The Airport Committee met with airport manager George Hoagland to review policies and issues. Dean John Flower underscored that the flight instructors were recognized as faculty and not "poor cousins" within the instructional program.[45]

KSU flight students gained a welcomed service on February 13, 1967, when the university's Campus Bus Service (CBS) began operations. Initially CBS service was limited to campus, but by August, off-campus routes were added, including one to the Stow-Kent Shopping Plaza.[46] Since the inception of collegiate flight training in 1939, students were required to find their own means of transportation to and from an airport for flying. Now, it was possible to board a bus on campus and ride out to the Stow-Kent Shopping Plaza. The walk along Kent Road to Paton Field was then less than a mile, although without the benefit of sidewalks and with exposure to all kinds of weather.

In February 1968, most likely due to the request of students or the Aerospace Technology Division, the West Main bus route was extended, providing thirteen trips a day to Paton Field. This modest service for airport classes contrasted with the forty-five trips per day to aid students going to shop at Murphy's Mart on the east side of Kent. By 1993, the West Main bus serviced the airport by request only.[47]

Leonard Mack checking oil on the Piper PA-23 Aztec, 1968 (Collection of Karen A. Mack and Michael R. Mack)

A huge setback hit the flight program in the summer of 1969. Leonard Mack, chief flight instructor, died suddenly at the age of fifty-two. Mack was a highly regarded career flight instructor, whose responsibilities since coming to Kent State in 1965 included being the executive pilot and designated FAA flight examiner for student checkrides. With his loss, instructor James Smith took over the executive pilot position, and William H. Wilgus was brought on as a temporary FAA-designated examiner.[48]

In 1969, the flight-training program continued to build, adding the flight courses necessary to meet the requirements of a professional pilot degree. With the new two-part professional pilot course, a student would first earn a multiengine rating and then have a copilot internship. The internship would provide the student twenty hours of flight time in the KSU executive airplane, a Piper Aztec. However, the flight program did not have a multiengine airplane to use for training. Otterson, program director, asked President White for approval to use the Piper Aztec so long as it did not interfere with university transport operations. White wanted experienced copilots only and felt that the solution would be to buy a second, cheaper Piper Apache when funds were available.[49]

Kent State had a bit of good fortune in February 1970 when James Armington donated a multiengine airplane. Armington was president of Buckeye Sports Center in Cuyahoga Falls and a former KSU student. The Beechcraft D18S was a similar model to the Beechcraft C-45 previously owned by the university but with one important difference—it was flyable. The airplane could carry seven passengers and be expanded to seat ten. This came at a perfect time; the new Beechcraft could replace the Piper Aztec as a university transport and free that airplane for the professional pilot course.[50]

In 1970, the FAA granted Kent State its own examining authority for the granting of the private pilot license. Previously, the final checkride for a student to earn the private pilot license had to be administered by an examiner from the FAA. Now, Wilgus had authority to perform the final check-

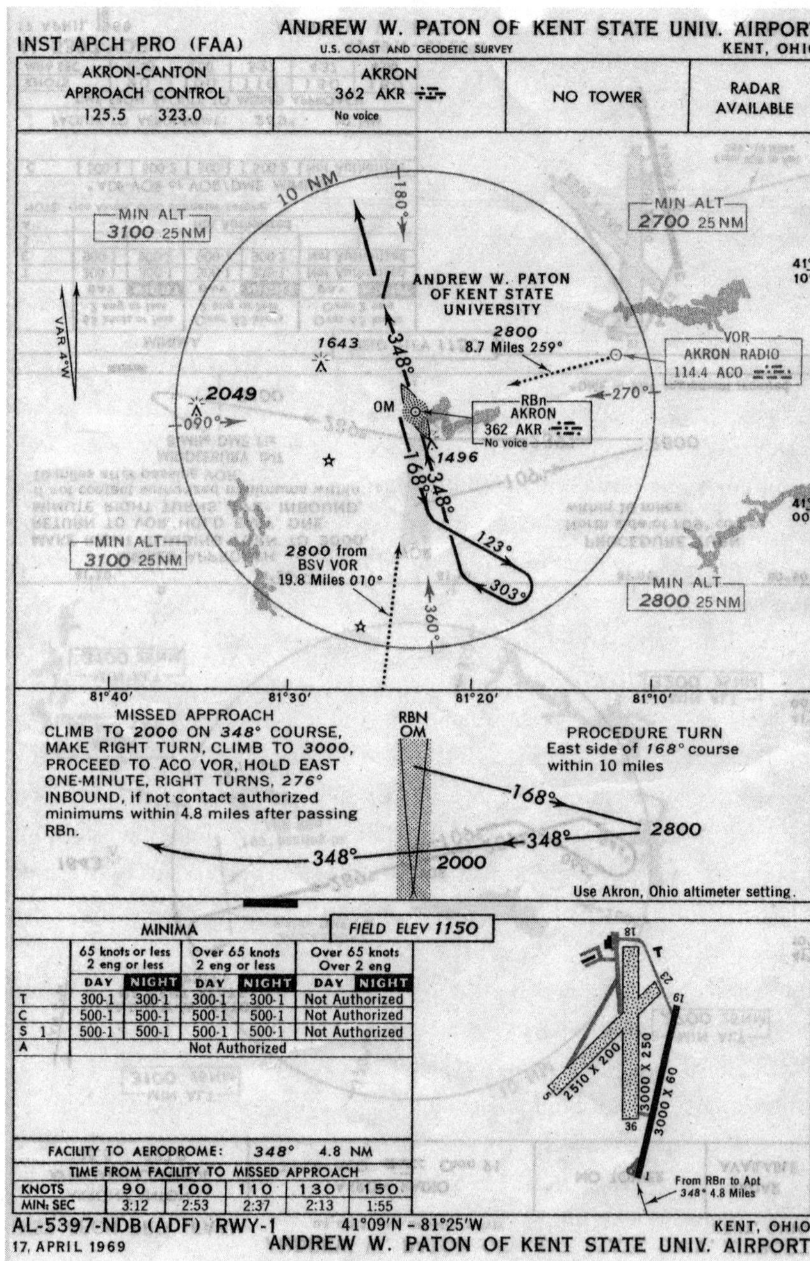

Non-Directional Beacon (NDB) Approach Chart for Andrew W. Paton of Kent State University Airport, US Coast and Geodetic Survey, 1969 (Ben Satyshur Collection)

ride and issue the license upon its successful completion. The authority was granted because of the quality of the KSU flight program, making it one of only a few in the country to been given this approval by the FAA.[51]

The May 4, 1970, Shootings

The tragic shootings at Kent State on May 4, 1970, received worldwide attention, placing the university in the history of antiwar protests of the period. In the coming years, the university would struggle to regain its footing and momentum. As a significant asset of the university, the airport played a role during the days before and after the shootings.

President Richard Nixon's invasion of Cambodia in late April 1970 triggered campus protests across the United States. On the KSU campus, a demonstration on May 1 dispersed without incident, but a disturbance that night in downtown Kent prompted Kent mayor LeRoy Satrom to declare a state of emergency. This was followed the next day by his formal request to Gov. James A. Rhodes for the Ohio National Guard to be sent to Kent. Arriving that night, the Guard encountered a large demonstration and the ROTC building aflame. Learning of these events, KSU president Robert White cut short his meeting in Iowa and asked for the university airplane to come to return him to campus. At midnight, the National Guard posted a guard at Paton Field.[52]

On the morning of May 3, Governor Rhodes helicoptered into Paton Field from a campaign appearance in Cleveland. Following a news conference in town, he waited at the airport for White's arrival and then left after a brief conversation to resume campaigning. That night, the Kent police received an unconfirmed report, determined to be a rumor, that four students had been arrested at Paton Field. Also reported was vandalism of six airplanes seven miles away at the Portage County Airport. Initial speculation, later dispelled, was that this was done in sympathy to events on campus. News of these reports at both the KSU and Portage County Airports reached Schwabe. Worried that something could happen at Paton Field, he went to stand guard.[53]

On May 4, 1970, approximately 2,000 students gathered on campus to protest. Attempts were made by the Ohio National Guard to disperse the crowd with no success. Tensions continued to rise, and at 12:24 P.M. guardsmen fired, killing four students and injuring nine. An hour later, the Guard placed a request for 400–500 rounds of tear gas. Captain Lars flew from Paton Field to Camp Perry in northwestern Ohio and returned at 4:45 P.M. with 1,200 rounds of tear gas. A two-and-one-half-ton truck was sent to the airport to retrieve the shipment.[54]

A total of seventeen aircraft were in place during the Ohio National Guard occupation, including two Cessna O1-A, two De Havilland U6A, and thirteen helicopters, some of which landed directly on campus. It is not clear what aircraft were stationed at the airport or what other locations were used as bases of operations. While the Guard did retain a presence at Paton Field through May 6, stories of bivouacking there cannot be confirmed and seem unlikely. Once the Guard departed, the Stow Police Department followed up for a time with a cruiser, making periodic checks.[55]

Flight-Training Program Develops

Eventually, normal airport operations resumed. Attention returned to establishing the new professional pilot course, specifically determining what airplane would be used for multiengine flight training. White gave final approval to use the Piper Aztec, and starting on July 13, 1970, it was used for student instruction. The donated Beechcraft D18S became the primary university transport with Smith as the pilot. Starting in fall quarter 1970, multiengine training was offered for the first time in the curriculum.[56] This would be the last piece needed to establish a four-year professional pilot degree program.

Between 1953 and 1970, the Kent State had seen the number of aerospace majors grow from 30 to 425, with 100 students taking flight training each year. Coursework was limited to two classrooms in Van Deusen Hall until a mezzanine was constructed in Room 111, providing individual offices for the staff and one additional classroom. The KSU flight program had been established, but plans for a new education building at the airport had been scrapped. Flight instructors had no alternative than to use the public lounge in the terminal as an instructional area. The most feasible short-term solution was to use a temporary portable trailer in the parking lot as a classroom. Already, the instrument flight simulator was mounted in a semitrailer at the airport. These classroom limitations threatened to restrict future enrollments, which could derail any new degree program.[57]

The winter of 1971 saw a new flight-training program offered to students. In a response to harsh winter conditions negatively impacting students' ability to fly, Kent State contracted with Page Field in Fort Myers, Florida, to provide flight training. Kent State sent six students and one flight instructor to Florida, where the students completed their private pilot license in five weeks. Plans were made to repeat the program the next winter with twenty students and two instructors participating.[58]

Leonard Mack and flight instructor son Michael with the Curtiss-Wright Dehmel Flight Simulator, ca. 1968 (Collection of Karen A. Mack and Michael R. Mack)

In May 1971, the School of Technology announced they would split the current aerospace technology degree into three new four-year degree programs. Originally outlined in the Airport Special Committee's 1964 report, the degrees would better prepare graduates to enter the progressively competitive job market. The new degrees were aerospace engineering technology, aerospace management technology, and aerospace flight technology. The aerospace flight technology degree would give the graduates the pilot certificates and training they needed to become professional pilots.[59] Previously, pilots had to supplement the training available at Kent State in other ways. For the first time, a student could come to Kent State University with the specific goal of becoming a professional pilot.

Summary

Robert I. White retired in September 1971 after eight years as KSU president, having faced many problems, including the tragedy of May 4, 1970. He is to be credited for seeing early on that the KSU Airport had greater education potential than was being realized. By creating the Airport Special Committee, he set the course for development of the airport and the flight program as Andrew W. Paton had envisioned it. The long-awaited paving of the runway and installation of runway lighting had finally occurred. What was created during White's tenure would last and set the course that has led to the present. Economic troubles for both the university and nation would be the significant factor in decision-making for the foreseeable future.

Chapter 10

Seeking Financial Stability, 1971–82

Kent State University's seventh and eighth presidents, beginning with Glenn A. Olds in 1971 and Brage Golding in 1977, grappled with dire university finances that hampered investment in the airport. The nation was in the throes of high unemployment and rising inflation. Ohio was losing jobs and suffering from inflation, stagnating wages, and rising oil prices. When Olds began, Kent State was experiencing repercussions from the May 4 event. Enrollment was down, and the university projected a budget shortfall of $1.3 to $3.5 million.[1]

Olds judged early on that Paton Field could be used more fully but learned it had a serious financial problem. As an auxiliary service, the airport had to be self-supporting, and, as Olds took office, it was running a $60,600 deficit. Proposed tuition increases led to student criticism of Olds for using a university airplane to fly to Cleveland. A letter to the editor in the student newspaper suggested it was time to sell the university transport airplane, the airport, the golf course, and the football stadium. Paton Field had its supporters though. Twenty aerospace students organized to advocate for the airport and how it might be made into a "self-supporting, profitable, and educational operation for Kent State University and its community."[2]

To address the revenue shortfall as well as the need to improve facilities, various proposals were made over the next two years to reorganize airport administration. At the same time, Kent State explored avenues for outside funding.

Reorganization Plans

The Aerospace Technology Division suggested the first reorganization plan, believing the very viability of Paton Field was at stake. The division proposed taking over operation of the airport, running it as an educational activity, and providing air transport service for university personnel. Cost savings of $36,000 a year would be achieved by discontinuing the use of part-time flight instructors and replacing them with Federal Aviation Administration–qualified students working as paid interns. The plan projected that this personnel change would balance the operating budget but at the same time acknowledged there would be no funds for capital improvements or unanticipated maintenance. Olds found the proposal offered "some dramatic changes in the organization structure and purpose of the airport facilities." Olds agreed that the Aerospace Technology Division would operate Paton Field as a trial for the 1972–73 academic year with a review to follow.[3]

To boost revenue and take advantage of a fairly quiet time in the summer, students, faculty, and staff were invited to take flying lessons, and qualified KSU students and staff could rent training airplanes when not in use. Finances began improving in fall 1972. To update the fleet, four airplanes were traded in as down payments on four new Cessna 150s and one Cessna 172 Skyhawk. By December, there were five Cessna 150s, two Cessna 172s, and a twin-engine, six-passenger Piper Aztec for university transport and instruction.[4]

Halfway through the trial period, Aerospace Technology Division director Peder Otterson reported to Olds that his unit's status as a division within an academic unit hindered its needed interaction with nonacademic depart-

Kent 2 (in 1974), one of four Cessna 150s bought by Kent State in 1972 (Kent State University Libraries. Special Collections and Archives)

Flight student Dave Boch planning a cross-country flight, 1977 (Kent State University Libraries. Special Collections and Archives)

ments and upper-level administrators. Becoming a department or school would facilitate communication and decision-making. Despite this impediment, Otterson reported that the academic staff wished to continue to operate the airport. Ralph Hetzel, dean of the College of Fine and Professional Arts, commended the aerospace technology faculty on their work but felt it required staff to carry an overload that was unsustainable. He also felt that the time required took away from academic responsibilities that neither the school nor the college could afford to give.[5]

As a follow-up to Hetzel's comments, the Aerospace Technology Division offered a second reorganization plan in April 1973. Otterson took exception to Hetzel's view that operation of Paton Field should not be within the academic unit and restated his faculty's concern about the future status of the airport. This new proposal was a long-range development plan that would make the airport "a financial asset as well as an educational and research center."[6]

The plan was to separate the Aerospace Technology Division from the School of Technology and place it in a new School of Aviation and Transportation with split reporting lines. For its academic programs, the new school would report to the vice president and provost for academic affairs. For airport operations and the university's air transport service, the school would report to the vice president for administration in cooperation with the University Foundation and the vice president for public affairs and development. Support for the general aviation community would be handled by contracting with a fixed-base operator, generally referred to as an FBO (a contractor with the right to operate at an airport and provide services such as fueling, hangaring, rentals, maintenance).[7] Albert Beckwith already operated the Commercial Aviation Corporation at Paton Field, serving general aviation, but this proposal suggested a formal contractual agreement.

Hetzel responded that responsibility for Paton Field should be assumed by a vice president on the nonacademic side of Kent State. If an aerospace technology faculty member were used as a manager, that time would not be charged to the department. Also, this was not the time to consider creating a new school. The decision to phase out the university's air transport service had already been made. Provost Bernard Hall agreed with this course of action, having previously recommended that Kent State contract with an FBO to operate the airport in its entirety.[8]

In June 1973, Olds decided the university would run Paton Field with a manager who would report to the vice president for administration. However, certain operations, such as maintenance, would be leased out. Otterson remained confident that having the airport run by the aerospace academic unit was still a sound plan. The Aerospace Technology Division successfully managed the airport's budget for 1972–73, generating a gross profit of nearly $41,000. If the unit had the necessary autonomy to make decisions and institute change, Otterson felt confident even more could have been accomplished. He reiterated to Olds the importance of Paton Field to the academic program and the division's interest in its development.[9]

The Tri-City Plan

While this discussion was going on, another proposal came forward in February 1973. It was drafted by William H. Wilgus, who was a Federal Aviation Administration (FAA) examiner prior to becoming airport manager in fall 1972. His plan reconceptualized the airport as a joint university/community operation. Wilgus believed Paton Field's survival depended on ensuring that necessary zoning, land acquisition, and financing would take place. His "master plan" envisioned a tri-city university airport under the direction and control of a port authority. The agency would represent Stow, Kent, Munroe Falls, and possibly Cuyahoga Falls.[10]

Such a plan would make Paton Field eligible for federal funding for needed improvements and would ensure that it was developed in a way that was complementary to Kent State and its surrounding communities. This would move the airport from a university-only facility to one with shared control.

At the time, the land surrounding the airport was largely undeveloped, but that was expected to change once local sewage problems were resolved and the state ban on sewer tap-ins in the area was lifted. Stow's development director James Farmer saw the airport as beneficial for attracting new business. With plans for Paton Field's expansion, he indicated that Stow could rezone the area to the east and northwest of the airport for light industrial use, including putting a road through to connect Kent and North River Roads. Rezoning was contingent on expansion that would make Paton Field more usable.[11]

Otterson and Charles W. Keith, director of the School of Technology, made initial community contacts to gauge interest. The Stow Chamber of Commerce gave its endorsement for future airport development. Wilgus shared his plan with university administrators in February 1973, but by April there still was no response. Frustration with the administration's apparent lack of interest led Wilgus to resign and take a position as aviation coordinator at a community college in Michigan.[12] Eugene G. Ripple became airport manager in 1973 and served until 1992.

Students, members of the community, and area leaders also noted the lack of university response to the tri-city plan. KSU senior Jeffrey Jack said students feared the airport would be shut down in June. The rumor of Paton Field closing caused John A. Doyle, manager of Akron-Canton Regional Airport, to write Olds in support of it remaining open. Concerned aerospace students sent Olds a letter opposing the closing with ten pages of signatures. Olds made no public response but did reply to the head of the student group that there was no plan to close the airport. Rather, he said, discussions centered only on phasing out the university air transport service.[13]

Options for Addressing Facility Needs

In addition to airport reorganization proposals, Olds had received reports of Paton Field's deteriorating and inadequate facilities from the very start of his administration. Similar to his predecessors, he had no funds to draw upon. Hetzel advised Provost Hall that before long he would need substantial expenditures for maintenance and capital improvements to keep the airport viable. The budget plan had called for the sale of both the Lockheed Lodestar and Beechcraft airplanes with the funds then being placed in an account for future airport improvements. Olds rejected that proposal and told KSU associate provost John Flower that the money needed to be placed into the general fund to "ease our extremely tight budget situation."[14]

Lockheed Model 18 Lodestar at the KSU Airport. The airplane was part of the Royal New Zealand Air Force during World War II and owned later by the B.F. Goodrich Company, which donated it to Kent State University in the late 1960s (Kent State University Airport)

As a cost-saving measure, free help was sought whenever possible. By fall 1972, flight students had remodeled and painted the interior of the small terminal building. Otterson was hoping to get Humble Oil Company to paint the sign on the front of the big hangar in advance of the upcoming intercollegiate flying meet. After receiving what he interpreted as a mild rebuke for not getting the rusting hangar painted using student help, Otterson replied to Olds that he felt it was too dangerous a job for students to tackle, given the lack of professional equipment and experience on large structures.[15]

In December 1972, Otterson sent a memo to Daniel L. Newcomb, University Foundation director, hoping that the foundation might be able to offer financial support to address the airport's needs. He prefaced his comments by stating that past university officials had neglected the airport and failed to realize its potential and that academic needs had also been ignored. The hangar's "true condition is not represented by peeling paint, but by the gaping holes through the roof." The total airport operation was supported by the income it generated, but that amount was insufficient to repair buildings or deteriorating runways. The most critical needs for Paton Field were resealing the runway and other blacktop surfaces and patching and painting the large hangar. For the academic program, Otterson sought a temporary classroom, specifically a mobile home, and annual replacement of Cessna 150 and 172 training airplanes.[16]

By February 1973, the condition of Paton Field's buildings had worsened. Otterson sent a memo to Hetzel calling it a "questionable safety hazard" and recommending that Kent State request an emergency appropriation from the Ohio General Assembly for renovation and repairs.[17] Given that help from the state was unlikely, other funding sources were sought to address the deteriorating airport facilities and the lack of instructional space. A proposal put forward by John Phillips, aerospace technology faculty member, was for construction of a geodesic dome to replace the large hangar. American architect Buckminster Fuller had visited the campus in fall 1971 and proposed that students build a dome. Architecture students did just that, constructing a dome between Taylor Hall and Memorial Gym. The proposed airport dome would be multilevel and accommodate classrooms and related service facilities. It would serve as an eye-catching landmark and have publicity value for the university. In a memo to Olds, Otterson commented that "with translucent panels it would shine like a cut gem and be visible for miles to thousands of people." Olds noted that it was a good idea, if funds were available.[18]

Fay R. Biles, vice president for public affairs and development, reported to Olds on several funding possibilities. One prospect was a Cleveland businessman who expressed interest in establishing a fixed-base operation and investing one million dollars in new buildings and runways. Environmental Structures, Inc., in Cleveland, offered a proposal for an air-supported hangar that would offer rental space.[19] Two other individuals indicated interest in donating T-hangars to the university and then leasing them back. One of these was Albert Beckwith, who in addition to his established business at Paton Field was also manufacturing manager at Amoco Chemicals Corporation. Beckwith would become an important fixture at the airport. Another promising lead was federal funding through the FAA. The FAA process required a master plan and matching funds. Biles felt Ohio State University's experience could inform them about handling the match.[20]

In May 1973, Newcomb reported to Olds that his contacts included William Zahuranic, lessee of Portage County Airport. Zahuranic was interested either in Kent State selling the current facility and constructing a new one in Portage County or leasing its current facility to him to operate. A year later, Portage County Airport officials urged Kent State to use their facilities, given that their airport was in a sparsely populated area and that increasing land values in Stow would make it difficult for the Paton Field to survive there.[21] This suggestion would surface again some years later.

Decision to Pursue FAA Funding

The university chose to pursue federal funding through the FAA for improvements to Paton Field. In September 1973, Olds wrote the FAA that Kent State intended to submit a formal proposal to undertake a planning study of the airport. The planning study would take nine to twelve months to complete and cost up to $45,000, with the university's share covered by in-kind services. Ripple, the new airport manager, expected the study to include consideration of forming a regional airport authority and the development of industrial parks. The board of trustees accepted the FAA planning grant agreement for developing a master plan on July 17, 1975.[22]

The resulting master plan was released in October 1976. It outlined improvements worth $4 million to be made over a twenty-year period. The federal government would provide $2.6 million and Kent State the remainder. The university's match could be made from a combination of a state grant, revenue from hangar rental, or reimbursement from earlier property purchases. With improvements, the number of airplanes based there was forecast to double and the annual number of airplane operations (a landing, takeoff, or touch-and-go by an airplane) to increase from 85,000 to 210,000 in twenty years. The recommendations that could be funded by the grant included a thousand-foot extension of the north-south runway, construction of a new east-west crosswind runway, and land purchase for clear zones. Other recommendations not eligible for federal funding included T-hangars, fuel tank and surface pumps, a training center, and new hangars.[23]

At hearings in summer 1977, citizens raised concerns that the planned airport expansion would increase noise and the risk of crashes. The proposed rerouting of North River Road to accommodate lengthening the main runway was roundly opposed. The alternate proposal called for lowering the road and building a runway overpass. Those voicing support for the plan argued there would be minimal safety and environmental impact and included the possibility of spurring commercial growth.[24]

Up to this point, the airport had had little success attracting business activity beyond the occasional corporate flight to bring personnel to a nearby facility. Paton Field was the closest field to several music venues, allowing bands to land closer than any large commercial airport allowed. A notorious incident at the airport occurred in 1975 when, after performing at the KSU campus, an airplane carrying six members of Joe Walsh's band ended up in a snowbank off the runway. During takeoff, the pilot discovered trouble with the controls and aborted the takeoff. It was later discovered that a whiskey bottle had become lodged in the rudder pedals.[25]

The FAA master plan was unanimously approved by the Stow Planning Commission in January 1978. Following all necessary review and approval by the board of trustees, the plan was submitted to the FAA. The FAA approved four million dollars in improvements in 1980, but federal funding issues would delay execution of the plan until 1982.[26] Because the acceptance of FAA funding obligated Kent State to maintain Paton Field into the future, this decision would be key to the airport's long-term survival and would thwart attempts by later university presidents to end university ownership.

Balancing Academic and Nonacademic Needs

Over the years, the KSU Airport was regularly singled out as a nonacademic extravagance. This criticism arose again in fall 1973 when the airplane fleet was updated. Five Cessna 150s and two Cessna 172s replaced airplanes that were sold and one whose lease was canceled. The balance of the cost was handled by flight fees. Nonetheless, some on campus criticized the purchase, suggesting that Paton Field was a frill and that the new airplanes were purchased through administrative deception from state funds, which prompted Otterson to write to the Faculty Senate Executive Committee. He wished to clarify that the airport was an auxiliary enterprise that supported the academic program of the Aerospace Technology Division. No funding came from the general instructional fund to purchase airplanes. Flight student fees paid airplane operating costs, plus depreciation and replacement, and the majority of instructors' costs.[27]

Air Force surplus F-84G Thunderjet acquired by Kent State in 1957 and located behind Van Deusen Hall, along with a Sikorsky helicopter donated by the Ohio National Guard, ca. 1978 (Kent State University Libraries. Special Collections and Archives)

As suggested in Otterson's 1973 reorganization proposal, the university closed the maintenance shop at the airport and leased the operation to V/STOL, Inc., creating the university's first FBO under its control of the airport. In July 1974, Kent State established a contract for anyone providing services for hire such as airplane rentals, flight instruction, charters, or air cargo. The airport would realize additional income from the sale of fuel, hangar space, and tie-down areas. Beckwith was the first to sign a contract for his Commercial Aviation Corporation in 1975, formally becoming an FBO at Paton Field. At the same time, Mensching Aviation of Akron became the new lessee for the airport's maintenance shop, providing service to local aircraft owners.[28]

The university held a meeting in May 1974 with users of Paton Field to inform them of increases in rent and services and to invite their comments and suggestions. As a result of that meeting, a booster club was formed to promote fund-raising activities for special projects. This goodwill for the airport would be tapped in the years ahead as it faced challenges to its survival. At the same time, concern arose that spring when AFROTC (Air Force Reserve Officer Training Corps) notified the university that due to low enrollment and a change in criteria, the KSU unit was no longer viable—a possible lingering effect of the May 4, 1970, shootings. Olds responded that the university would "bend every effort" to return the detachment to viability, and by October enrollment exceeded the required target figure.[29]

The end of the 1975 and 1976 fiscal years brought welcome news that the airport's budget showed a small surplus. In addition, Kent State included the airport in its capital request to the board of regents, specifying the need for $2.3 million for airfield renovation and expansion and $730,000 for an aerospace technology building.[30] Although the request was not funded, the Aerospace Technology Division faculty was gratified that the university acknowledged the airport's needs.

Improvements Wait for FAA Funding

Olds gave personal attention to many airport issues upon becoming president in 1971. He chose to get firsthand experience with the operation by taking flying lessons in the summer of 1973. After enrolling as a student, he registered for Aerospace 270 and had a senior aerospace major as his instructor. He was expected to earn his private pilot license by the end of summer, but the outcome was not reported. Olds submitted his resignation in November 1976, effective at the end of the academic year. He would still face big challenges in the last months of his presidency, including an energy crisis, effects of a poor statewide economy, and a collective bargaining impasse. The Ohio General Assembly passed a resolution noting Olds's dedication and service to the university, particularly his handling of the "trouble-ridden years following May 4, 1970."[31] During his tenure, he explored a number of solutions to address the airport's financial problems and capital improvement needs, including reorganization and outside financing. Significantly, the decision to seek FAA funding would become a major part of his legacy with regard to Paton Field in the years ahead.

Old's successor, Brage Golding, became Kent State's president in September 1977. He was charged by the board of trustees to bring financial stability to the university. State revenues were down, causing a reduction in subsidy to higher education. The university had drawn from reserves and delayed maintenance to pay legal bills and security costs related to May 4. The tragedy also led to serious enrollment problems over the period of 1970–78. Golding reorganized his administration, cutting three of the six vice president positions. With the elimination of the vice president for administration, the airport would now report to Richard E. Dunn, the vice president for finance and treasurer.[32]

During Golding's five-year term with its many staff reductions and operational cuts, the airport would be relatively untouched. With the decision to close University School, the university's elementary school considered by many to be "a semi-private elite school," parents questioned why operations such as the airport and golf course were not also considered for cutbacks. Golding explained that as auxiliary operations, they had to pay their own way.[33] For the KSU Airport, revenue came from flight fees, airplane rentals, and sale of fuel.

Contrary to the campus at large, the School of Technology reported increased enrollment for its majors in 1978–79. The change from a quarter to a semester system began in fall 1979. Enrollment in flight classes doubled over the previous term, and the number of certified flight instructors was increased from eleven to eighteen. Airport manager Eugene Ripple commented that the program's reputation was enhanced by taking in more students "as long as we're safe. As long as our instructors are good." Importantly, additional airplanes were obtained to accommodate the increased number of flight students. Most were leased. Going into fall semester 1981, there were twenty-five university airplanes.[34]

In February 1982, three new flight simulators were installed at Paton Field, at a cost of $12,000 each. Not only did they enable students to continue their flight training despite poor weather, but they also provided an effective means to reinforce proper procedures. A mobile trailer housed the simulators and provided a classroom and space for briefing and debriefing. Flight instructors previously had given ground instruction in the terminal lobby, using the lobby windows as a whiteboard for instruction.[35]

Flight instructor group photo, 1981. *Left to right:* unidentified person, Kirk Kubias, Ron Thompkins, Gary Landrum, unidentified person, Ken Lazor, Lyn Carlson; on the wing: Darryl Lupo, Brent Turner, Richard Schwabe, and Bill Miles. (Kent State University Libraries. Special Collections and Archives)

Trailer for the FAA Examination Center, 1981, which later was still referred to as the "FAA trailer," although FAA testing was moved to campus in 1996 (Kent State University Libraries. Special Collections and Archives)

Despite strong enrollment in the flight program, airport revenue was insufficient to undertake major improvements. The hope for capital improvements lay with the master plan, which had been submitted to the FAA in May 1977. Even when approved by the FAA in 1980, funding awaited the needed congressional appropriation. The FAA would pay 90 percent of the four-million-dollar grant, and the remainder would come from Kent State or the state. The grant, however, could not be used for buildings. The need for improved and increased space for classrooms and terminal space was acute; Ripple said simply, "the physical facilities are rotten."[36]

Ripple prepared a plan that included new offices, a weather room, flight-planning rooms, and space for FAA testing with the hope for future local and state funding. FAA written testing had required students to drive to the Cleveland Flight Standards District Office (FSDO). Then testing was moved to Dix Stadium on the KSU campus. FAA inspectors would come to Kent State and administer the test to all requiring the written exam in the stadium press box. In 1981, a mobile trailer was placed at Paton Field to serve as the FAA trailer. Barb Ripple, Eugene Ripple's wife, was the first FAA test administrator at the airport.[37]

Some needs required immediate attention. In December 1980, the maintenance hangar roof was declared unsafe. The board of trustees determined that since the entire roof needed to be replaced, it would be both economical and the ideal time to construct a necessary addition to the building. The costs were estimated to be $60,000 for a concrete-block addition and $52,000 for the new roof. These would be covered by airport income over a two-year period. Some issues, such as the poor condition of the taxiway, could not

Above: Aerial photo shows the two grass runways and the one paved runway connected to the ramp by a single paved taxiway, 1981. Cinder taxiway connects to the south end of the runway and its paved turnaround. (Kent State University Libraries. Special Collections and Archives)

Left: Office trailer for Commercial Aviation Corporation, the airport's fixed-base operator, 2016. It was owned and operated by Albert Beckwith. (Courtesy of C&S Engineers, Inc.)

be addressed. The taxiway was a mixture of sod and gravel and was rarely used due to its poor condition. Pilots landing to the south would roll to the end and wait adjacent to the runway until it was clear to taxi back north to the ramp and hangar.[38]

Extracurricular Flying

Throughout the 1970s, the KSU Flying Club experienced new life after the downturn it suffered when the university no longer needed its members for flight instruction. The club resolved its financial problems, and by 1972 recruiting efforts had brought the membership up to twenty-five. Its status changed from being an independent group to one limited to licensed students and KSU employees. As such, it partnered with the university to purchase a Cessna 172 and gained access to other university airplanes at off-times.[39]

KSU Flying Club members, joined by some aerospace technology majors, formed a flying team to represent Kent State in the regional and national competitions sponsored by the National Intercollegiate Flying Association (NIFA). The KSU Flying Club's team had great success at the Great Lakes regional meets, which it hosted at the airport in 1972 and 1978. The team qualified to go on to the NIFA National Safety and Flight Evaluation Conference (SAFECON) meets in 1972, 1973, and 1980. By 1982, the flying team was no longer a KSU Flying Club activity but had become the Precision Flight Team under the direction of the flight program.[40]

First Annual Air Show

Another welcomed feature of the 1970s was the creation of an annual air show. It began as part of a three-day aerospace exhibition in May 1973, called Air Expo '73. The first two days were held on campus and featured a trade show, seminars, and exhibits, including an Apollo 11 moon rock. The air show was on the third day and featured skydivers, aerobatics, and aircraft demonstrations. This began a long history of air shows at Paton Field that would be held annually under the Air Expo name through 1982, most often sponsored

The KSU Flying Club flying team following the National Intercollegiate Flying Association meet, June 19, 1980. *Left to right:* Tom Matovich, Darryl Lupo, Mike Hjort, Jay Webber, and Mark Madden. *Back:* Lyn Carlson, Art Ritchie, and Bill Miles. (Courtesy of the *Daily Kent Stater*)

Above left: Ruth Sitler at the first Air Expo, May 15, 1973 (Kent State University Libraries. Special Collections and Archives)

Above right: Air Expo '82 program (Ted Orris Collection)

Left: Student propping the T-6 Texan used by Richard Schwabe at Air Expo '83 (Scott Mills Collection)

by the KSU chapter of the American Institute of Aeronautics and Astronautics, the KSU Flying Club, and the professional aviation fraternity Alpha Eta Rho. The Air Expo became a well-attended annual event that drew in the community and highlighted the flight program and the airport. During Golding's presidency, the expo aligned nicely with his marketing campaign designed to create a more positive profile for Kent State and to recruit more students.[41]

The traditional pancake breakfast provided by the KSU Flying Club began with Air Expo '74. The typical air show included hot air balloons, static displays, aerobatics, and airplane rides. Some years had unique offerings. Airplane rides were offered at Air Expo '74 at the cost of two cents per passenger-pound. Air Expo '76 included a contest for Ms. Air Expo; the winner would receive a free airplane ride. Air Expo '77 sold beer for thirty-five cents a glass or three glasses for a dollar.[42]

Attendees at Air Expo '79 gather on a light twin-engine airplane to watch the show (Kent State University Libraries. Special Collections and Archives)

In addition to static displays and rides, the Air Expo often had special events, such as wing-walking (1978), car-to-airplane transfer (1979), and parachutists (1979, 1980). A flyby performed by two Air National Guard A-37s always thrilled the crowd (1978–80). The 1982 show featured a P-51 Mustang, a C-130, and a P-3 Orion. Richard Schwabe, now a full-time FAA designated examiner, regularly performed acrobatics in a World War II T-6 "Texan" trainer.[43]

This student-run event ended with Air Expo '82 as costs for booking aircraft and acquiring the necessary insurance continued to increase. The Air Expo had met its goal of promoting aviation and the aerospace program to the university community and the general public.[44] The next air show at the airport would not be held until 1995.

Summary

As Brage Golding's tenure as president came to a close in June 1982, the US recession had not yet ended. During Golding's administration, the KSU Airport escaped direct budgetary cuts because of its status as an auxiliary operation. Although enrollment in the flight program had been strong and the aircraft fleet had been updated, airport maintenance and capital needs persisted. Hope for Paton Field's future was centered on the FAA master plan submitted under KSU president Glenn Olds and approved in 1980 during Golding's presidency. The plan still awaited funding, which would not happen until yet another presidential administration was in place.

Chapter 11

First Improvements, Then a Threat, 1982–91

When Michael Schwartz followed Brage Golding as Kent State University president in 1982, finances were tight. Schwartz believed making decisions on which programs to retain and which to reduce or eliminate was hindered by the university's ill-defined identity. He developed a new mission statement that affirmed the primacy of a general liberal arts education and a secondary commitment for career preparation, signaling that professional training programs could be eliminated.[1] This emphasis on mission would become an important argument in considering the future for the flight program and Paton Field later in Schwartz's administration.

Long-Awaited Airport Improvements

Times were promising for the airport at the start of the Schwartz presidency as the airport would finally receive the federal funding based on the Federal Aviation Administration (FAA) master plan. The university first began pursuing FAA funding in 1973, undertook the master planning process in 1975, submitted it to the FAA in 1977, and obtained FAA approval in 1980. Now with funding, the plan would be executed in five phases over seven years. KSU Airport manager Eugene Ripple, who was instrumental in writing the master plan, would play a key role in its implementation. As he said at the outset, "It's the start of a whole new era [for the airport]."[2]

For each phase, the FAA contributed 90 percent of the funding with a required 10 percent match from Kent State. The match was variously handled by grants from the Ohio Department of Transportation, credit for previous land purchase, airport revenue, and other local sources. The five phases of the plan were as follows:

- Phase 1 (1982, $151,773) covered construction of a partial asphalt parallel taxiway.[3]
- Phase 2 (1982–83, $545,813) included taxiway completion, runway overlay, apron expansion, and land clearing and restoration.[4]
- Phase 3 (1983, $283,333) completed the taxiway and upgraded the runway lighting system and provided for purchase of fifteen acres of land north of Kent Road.[5]
- Phase 4 (1985, $439,000) purchased three properties on the south side of Kent Road for a clear zone.[6]
- Phase 5 (1987–88, $1,624,019) purchased twenty-four more acres of property north of Kent Road and extended the north-south runway to four thousand feet.[7]

Tree clearing was a big project in 1983 during Phase 2. To realize savings over what a professional business would charge, the crew consisted of eight students, four Amish men, and two horses. Their task over the course of the year was to clear thirty-eight acres of oak, maple, cherry, and aspen trees as well as underbrush. Felled timber was cut into lumber at a sawmill erected on the site and then sold.[8]

During Phase 3, the paved runway was shut down for the runway lighting upgrade during the latter part of 1983. Workers dug trenches on either side of the runway, and flight operations were relegated to the grass runways. The ground froze before the project was completed, and grass runway 5–23 was plowed during the winter to maintain operations.[9]

Breaking ground to mark the beginning of improvements funded by the Federal Aviation Administration master plan, starting with paving of the taxiway, August 1982. *Left to right:* College of Fine and Professional Arts dean Thomas Barber, business affairs vice president Richard Dunn, KSU airport manager Eugene Ripple, and KSU president Michael Schwartz. (Kent State University Libraries. Special Collections and Archives)

During Phase 4, in the summer of 1986, the three houses south of Kent Road had been moved and one business demolished. Two of the houses were relocated to Baird Road and the other further west on Kent Road.[10]

On June 1, 1987, at the Business Affairs/Institutional Advancement Committee meeting of the KSU Board of Trustees, Richard E. Dunn, vice president for business affairs, advised members that the initial plan for Phase 5 provided for extending the north-south runway to only 3,950 feet. Discussions were under way to

Tree clearing by the Amish and their horses, 1983 (Kent State University Libraries. Special Collections and Archives)

Lumber mill on site for tree-clearing project, 1983 (Kent State University Libraries. Special Collections and Archives)

negotiate for the extra space needed to provide the 4,000 feet required by the FAA to be designated a general utility airport. In December, and the university acquired five properties north of Kent Road for the needed clear zone and demolished three houses on the land.[11]

While the FAA capital improvements were under way, Kent State was faced with the deterioration of the main hangar's roof. In November 1983, an inspection found the corrugated steel sheets were rusting and could be blown off by the wind, so it was determined to be a safety hazard. The roof was removed down to the structural steel girders. Dunn said the strength

Right: Hangar roof removed due to safety concerns, winter 1983 (Kent State University Airport)

Below: Hangar reskinning project included preserving the original Van Devere block storage area, visible on the right side, spring 1984 (Kent State University Libraries. Special Collections and Archives)

of the flight program and its increasing enrollment necessitated that the facility be repaired or replaced. Over that winter, temporary tie-downs were installed inside the hangar to secure the airplanes, which were still pulled in each night although there was no roof.[12]

Drainage was an ongoing issue. In 1989, the airport received an aviation improvement grant of $97,200 from the state. Along with the university's match of $10,800, the funding was used to improve drainage in the hangar area and extend an apron.[13]

Airport manager Ripple sent a note to public customers of Paton Field in March 1984, updating them on the improvements and advising them of preliminary discussions for a new terminal building should capital funds become available. In that note and again a year later, he asked that anyone interested in building T-hangars at the airport under a land lease contract contact him. Albert Beckwith of Commercial Aviation Corporation, responded both times that he wished to pursue the possibility. Ripple also thanked customers for their patience regarding higher prices and poor facilities, stating his belief that "those days are finally past and the future of our airport is bright both for the immediate future and the long-range plans."[14] No T-hangar project developed, even though the extension of the north-south runway was expected to make Paton Field more attractive to corporations and private pilots for airplane storage, but that required adequate hangar space.[15]

Problems Follow Program Growth

Dunn's earlier comments to the board of trustees about the strength of the flight program were echoed by Ron Price, chief flight instructor, in January 1984. Price reported 173 students had enrolled, an increase of forty over four years, due to the willingness of Kent State to allow the program to grow. Some concerns did surface in fall 1984. A group of concerned alumni wrote Schwartz about what they perceived as a decline in the quality of the program over the past five years when it "was expanded without competent supervision and control." However, several weeks later, the FAA notified Schwartz that their renewal inspection found the KSU pilot training school in compliance with FAA regulations (FAR 141). Still, given that four to five incidents had occurred over the past year, they recommended additional supervisory personnel be in place to perform and monitor training, to provide more "direct supervision."[16]

David Mohan, director of the School of Technology, promptly wrote Terry Roark, vice president for academic and student affairs, heralding the long-standing reputation of the program for graduating highly skilled pilots and forecasting continuing growth in admissions. Paton Field was central to the program's success and its growth. He asked for patience as the concerns over supervision, safety, budget, and priorities were addressed. His memo was sent on to Schwartz with Roark's endorsement and a suggestion to explore increasing general aviation revenue at the airport.[17]

Extensive changes were put in place. Price commented these transformed the program "from a small to a large operation with a big emphasis on safety."

Former students Dave Schrank and Terry Sasala visit with mentor Richard Schwabe, 1987. It was common for an alumnus to radio the KSU Airport dispatcher when flying high overhead and ask to talk with Schwabe. (Dave Schrank Collection)

Additions to the program included a two-way radio, a full-time dispatcher clerk, educational specialists, and a computer.[18] The airplane fleet was updated as well, including, from 1984 to 1987, one Piper Aztec, two Piper Seminoles, thirteen Cessna 152s, five Cessna 172 Skyhawks, and a number of Cessna 172RG Cutlass models.[19]

The KSU Precision Flight Team hosted the National Intercollegiate Flying Association (NIFA) Region III Safety and Flight Evaluation Conference (SAFECON) with competitors from five other schools in November 1982. The traditional events included precision landings, aircraft recognition, flight navigation, ground trainer, and bomb drop. The team's first place finish qualified them to compete in the national competition in May 1983. Paton Field again hosted the NIFA Region III SAFECON in October 1985; however, there were no air shows from 1983 through 1994, covering the entire period of the Schwartz presidency. Possibly this was due to the expense of putting on the community air show.[20]

The lack of adequate space at the airport for offices, classrooms, and the public remained acute. Ripple commented that "space is at a premium. . . . Some days you just can't move in the halls here." The FAA inspectors, as part of their regular recertification process in 1986, found the airport's ground training facilities to be in direct violation of FAA regulations and totally inadequate for the program. To create the needed space, Schwartz and Roark agreed to add a double-wide trailer. This would join the one already in place, which was outfitted with four flight simulators and eight cubicles and was used for classes, flight simulation, briefings, and debriefings.[21]

Early in the morning of April 7, 1986, Kent State's twin-engine Piper Aztec ran out of fuel and made an emergency off-airport landing at Arizona's Salt River Indian Reservation. The five occupants—three flight instructors and two students—were injured to varying degrees. Plane crash victims filed lawsuits against Commercial Aviation Corporation, which had performed maintenance on the airplane, as well as against Kent State. The university lawsuit was settled out of court in December 1988 for $909,000. This accident was serious in both its human and financial outcomes and would influence future events. One positive consequence was the incorporation of Standard Operating Procedures (SOPs) for flight training, which created more structured operational guidelines.[22]

Double-wide trailer added in 1986 to provide necessary classroom space. The adjacent trailer housed simulators and provided limited instructor space, 2017. (Photo by William D. Schloman)

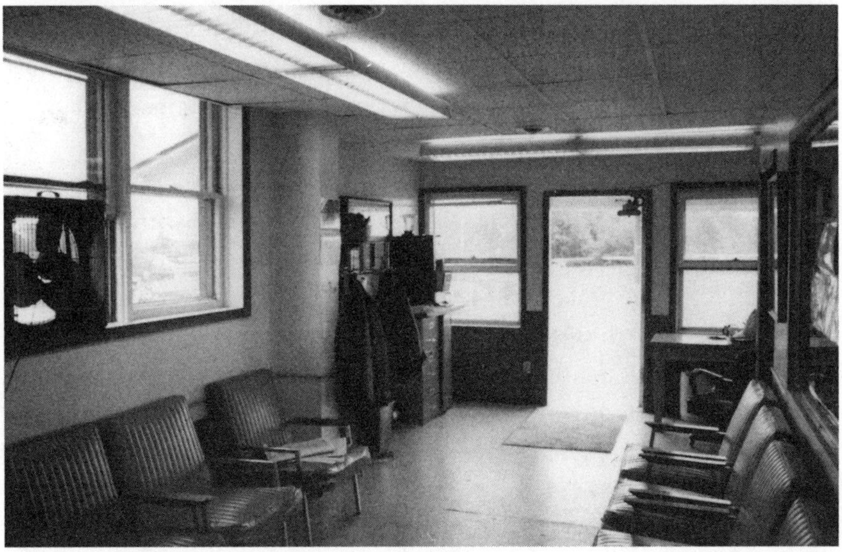

KSU Airport terminal lobby, early to mid-1980s (Kent State University Airport)

Pressure from Local Residents

With the completion of the runway extension, in 1989 a group of Stow and Munroe Falls residents voiced their concerns to Stow City Council about airport noise and safety. This was triggered by noise from a British Hawker business jet landing at the airport in 1988. About 250 citizens formed the Association for Sensible Airport Policy (ASAP) to pressure Kent State into addressing their concerns. The association began a petition drive demanding the flight pattern be raised, jet aircraft be banned, certain practice flights be ended, and airport expansion cease. Beckwith notified his contacts that their support of the airport was needed at an upcoming meeting of the Public Improvements Committee of Stow City Council on June 26, 1989. The full house heard arguments on both sides about Paton Field. The council proposed an ordinance that would give the city more control over the airport.[23]

A heated meeting in October between the Public Improvements Committee and residents resulted in a committee vote to recommend to council that all jets be banned at the airport. As a conciliatory gesture, Kent State agreed to raising the flight pattern from 614 to 866 feet, moving it southward over the Cuyahoga River, banning touch-and-go exercises during certain hours, prohibiting high-powered jets with unmodified engines, and cleaning up the clear zone north of Kent Road. A council member responded that the changes were not enough, which was countered by the mayor, who suggested that some council members were playing politics. Stow City Council received a petition signed by more than a thousand residents asking the city to ban all jets from Paton Field beginning January 1, 1990. An airport official had estimated there were two jet landings per month, and Beckwith pointed out that the university was working on a list of specific jet aircraft to ban because of undesirable noise levels.[24]

Owners of two residences to the immediate west of the new northern clear zone complained in 1989 that the noise levels had increased with the runway extension and that their hope for a sound attenuation mound had not materialized. The FAA proceeded to provide grants later that year for the university to purchase those two properties as well as an adjacent vacant lot.[25]

Although ASAP continued to pressure for more restrictions on Paton Field, the Stow law director had determined that legislation banning jets would be unenforceable. However, a Supreme Court decision in 1980 allowed municipalities to have zoning control over state-owned operations. The question then became whether jet activity that began after 1980 gave the city the authority to restrict or eliminate it. The new proposed legislation would require Kent State to apply for any changes at the airport that had occurred since 1980. When the legislation came up for a vote in February 1990, it was amended to require the university to submit any changes in progress beginning with January 1990. The legislation passed.[26]

More Program Problems Surface

Going into 1988, the KSU Airport was humming. Kent State had the fourth largest university flight school in the country based on flight activity. More than 200 students were taking flight courses with another 100 enrolled in the program. Ripple reported the airport had over 114,000 landings in 1987, mostly by students, making it the busiest small airport in the five-county

Akron area. There were four flight chiefs and seventy flight instructors. Given the facilities and the number of airplanes available, the program had frozen enrollment.[27]

Student complaints arose in May 1988 about the facilities and limited number of airplanes. The university then planned to restrict the program further by tightening admissions and reducing enrollment by about 25 percent. Looking ahead, Schwabe cautioned that it would soon be necessary to start updating the fleet of twenty-four airplanes, most of which would soon be five years old. There was total agreement that the facilities needed improvement. One student called the terminal a "run-down shack" and too small to handle the nearly seventy students needing to work in the area.[28]

In January 1988, Cheryl Casper, the associate provost for academic and student affairs, reviewed the financial history of the flight-training program for the three previous years and found an increasing operating deficit. In transmitting the report to Schwartz, Thomas Moore, provost and vice president for academic and student affairs, indicated his lack of support for maintaining the program and the airport.[29]

Moore's opinion must have been known within the flight program. Two aerospace faculty members proposed to Schwartz in November 1988 that they take over the flight training as private contractors. They suggested this would relieve Kent State of liability and provide management continuity and quality control. In May 1989, School of Technology director Mohan sent a memo to Thomas Barber, dean of the College of Fine and Professional Arts, recommending that the flight-training program be converted to management by a fixed-base operator (FBO). This would have meant returning to an operational environment similar to the one with Van Devere Aviation in the 1960s.[30]

At Moore's request, Casper again reviewed the status of flight training in June 1989. Enrollment figures showed significant growth in the program—from 140 students in 1978 to 385 in 1988. Her analysis of income generation found cost exceeding income for each of the years analyzed. The conclusion was that "if the program were to be eliminated or phased out, the drop or shifts in enrollment could likely be sustained by the university." She acknowledged that this was based on finances alone and that any decision would need to consider the broader question of institutional mission.[31]

Suspension of Flight Program Admissions

Whether it was primarily to refocus the mission or to deal with financial pressures, on October 26, 1989, President Schwartz announced to key staff

Wilbur Biggin and Tom Grossman speak with instructors following KSU president Michael Schwartz's announcement to halt flight program admissions, the *Daily Kent Stater,* November 8, 1989 (Kent State University Libraries. Special Collections and Archives)

his decision to suspend admissions to the flight program, initiate the process for program review and abandonment, and redirect the School of Technology programs into high technology. Richard Dunn was directed to review what was needed to close Paton Field. Moore made the public announcement on November 2 that admissions were being halted and a review of the program could possibly lead to its abolishment. Currently enrolled flight students would still have time to earn their degrees. Moore maintained that the central reason for the closure was that the flight program did not fit into a research agenda to advance technology; however, others saw the recent lawsuit settlement for the 1986 Aztec airplane crash and hassles with the City of Stow as contributing factors.[32]

Reaction from flight students, instructors, and parents was immediate. Several meetings were held on November 7. At an informational meeting with three hundred to four hundred students, Barber explained that flight training did not fit into a research institution, as it was vocational preparation, not academic training. Flight instructors met to discuss how the program might be saved. Tom Grossman, flight-training coordinator, felt that there were research possibilities with the program. Instructors decided to collect funds to wage their campaign to save the program. Later that night, a small group, including flight majors Wilbur Biggin and Scott Koeppl, met at the Loft, a Kent bar, and formed the Organization for the Continuance of Aviation Education (OCAE).[33]

The *Daily Kent Stater* ran a story on November 9 stating the KSU Airport had a reported deficit of $276,458 for the period July 1985 to July 1989. Ripple said maintenance costs were the main reason for the shortfall. Moore commented that although academic factors led to the decision to suspend admissions, finances were a consideration. Dunn offered that if the decision were made to eliminate the flight program, "my first choice would be to get rid of the airport somehow." Schwartz informed the board of trustees that a recommendation based on the review of the flight program would be presented to the board, hopefully in spring 1990.[34]

Following Schwartz's decision to suspend admissions to the flight program, Moore asked Mohan to provide an objective assessment of the flight-training program for those who would determine its future. Mohan's thirty-page report covered the issues of the deficit fund balance, liability, and program focus. Mohan acknowledged that the number of majors should have been capped at a lower figure to deal with capacity issues of instructors and airplanes. In any case, flight fees were an inadequate means for funding the program, making the likelihood of running a deficit a certainty. While Kent State's safety record at Paton Field was commendable, risk was always present—as seen with the 1986 Aztec accident in Arizona. He noted that the issues of funding and liability were what had led other universities to contract out their flight training to an FBO. He personally believed that if the School of Technology were to move away from the application of technical content toward a research focus, aviation education should not be continued. Mohan's report included several plans depending on what course was chosen. These included a phased plan for ending the program, considerations for using an FBO to manage flight training, and several options for upgrading the fleet.[35]

Beckwith proposed to Moore in December 1989 that Commercial Aviation Corporation assume total responsibility for the flight-training program, stressing the reduction in liability for the university. Beckwith met with Moore on January 4, 1990, and outlined his plan for a phased transition of the program. He believed that although his proposal did not include operating the airport, it would improve Paton Field as an asset for flight training and for the community. Moore indicated to the *Daily Kent Stater* that the whole role of the FBO was being reviewed.[36]

Following its founding on November 7, OCAE mobilized. Its members would include flight majors, flight instructors, and flight-training staff, faculty, and alumni. Led by Wilbur Biggin, the group's goal was to gather information, provide facts, and develop solutions that would allow the program to continue. Points included in its organizational preamble were that the flight training was no more or less vocational than other professional programs on campus and that this was an opportunity to help define its future role. The document included a code of conduct, which required acting in a professional manner, avoiding nonconstructive criticism and negative publicity, and working to be part of the solution. The group had officers, a faculty advisor, and chairs for its sixteen different committees. There was an active letter-writing committee to garner support from alumni, the aviation industry and organizations, and other aviation professionals. Fact-finding included inquiries to other flight programs. Fund-raising and publicity would help get the word out.[37]

An open letter from Biggin to Moore appeared in the *Daily Kent Stater* on November 30, pleading that the decision to halt program admissions be rescinded for fear that it would in and of itself result in the demise of the program. The letter was endorsed by the Undergraduate Student Senate, Inter-Fraternity Council, and Black United Students. The Undergraduate Student Senate had already gathered a thousand signatures calling for reinstatement of admissions. Schwartz and Moore responded in a joint letter sent to Biggin on December 1, outlining their argument that it would be unconscionable to both incoming students and Kent State to continue admissions to the flight program when its cancellation was a real possibility.[38]

A December 7 memo to Moore signed by School of Technology faculty echoed OCAE's concern about the ongoing impact the suspension of admissions would have and asked for time to prepare a proposal for consideration before a final decision was made. Moore responded to the faculty that it would be unwise and unfair to rescind the suspension of admissions. He also reminded them that he had raised the question of the role of the flight program two years earlier and that they should have been addressing the development of curricula to support new technologies.[39]

Schwartz almost immediately received letters from numerous political figures and airline executives. His response to Ohio senator John Glenn was similar to the others. The rationale for reviewing the flight-training program was to focus on the university's central mission, as well as to address budget shortfalls and liability concerns. "My own personal view is that program is not worth the cost or the risk." His letter to the United Airlines' vice president added that "we have never had any support for this program from the airline industry." Executives at American Airlines and TWA received the same message.[40]

In the spring, OCAE actively worked to contact flight alumni, seeking their support as well as donations to continue protesting the administration's plans. Over a two-week period in March, OCAE took out five different "Flight Training Fact" ads in the *Daily Kent Stater* that were addressed to Provost Moore. Each covered a different topic, including mission, flight student revenue, buildings, utilities, and state subsidy.[41] On March 13, OCAE held an open meeting to answer questions. The attendees included Wilbur Biggin, Scott Koeppl, and alumni Ted Orris and Dr. Gene Petrus.[42]

In January, David Dennis of the Ohio Bureau of Aviation related to Beckwith that both his agency and the FAA were strongly opposed to closing Paton Field. Kent State would face possible default and violation of grant agreements with both agencies. Similarly, they would also oppose the closing of the flight school, as its existence was the major reason for the airport grants. The *Daily Kent Stater* reported on April 20 that John Kinville, an FAA airport engineer, had met with Dunn as early as February to discuss the program and Paton Field. Kinville reiterated that the university was committed to operating the

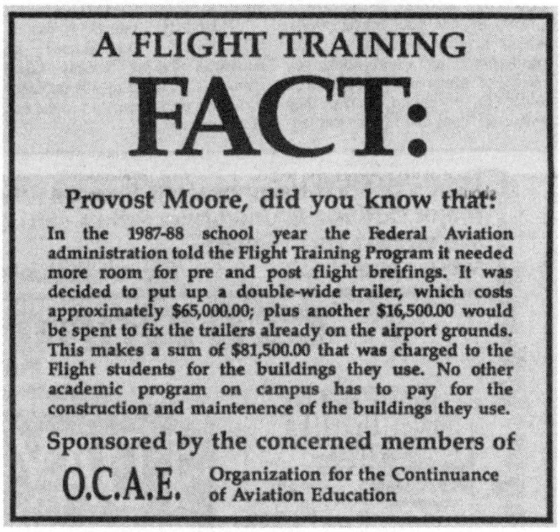
A FLIGHT TRAINING

FACT:

Provost Moore, did you know that:

In the 1987-88 school year the Federal Aviation administration told the Flight Training Program it needed more room for pre and post flight breifings. It was decided to put up a double-wide trailer, which costs approximately $65,000.00; plus another $16,500.00 would be spent to fix the trailers already on the airport grounds. This makes a sum of $81,500.00 that was charged to the Flight students for the buildings they use. No other academic program on campus has to pay for the construction and maintenence of the buildings they use.

Sponsored by the concerned members of

O.C.A.E. Organization for the Continuance of Aviation Education

Above left: Flight-training fact ad sponsored by the Organization for the Continuance of Aviation Education, the *Daily Kent Stater,* March 7, 1990 (Kent State University Libraries. Special Collections and Archives)

Above right: The Organization for the Continuance of Aviation Education open meeting at the Kiva with Scott Koeppl at the podium, March 13, 1990 (Scott and Heidi Koeppl Collection)

KSU Airport for twenty years, having accepted FAA funding. The only options were to pay back the funding or find a public sponsor to take over operations at Paton Field. The cost of a buyout was estimated at one million dollars. Plans went forward for the next board of trustees meeting on April 21 to include presentations on the future of the flight program by flight instructor and OCAE president Wilbur Biggin supporting the program and two area residents representing ASAP opposing it.[43]

Reinstatement of Admissions

Unexpectedly and with no fanfare, Schwartz issued a press release on May 1, 1990, reinstating admissions to the flight program beginning in the fall semester. To address the concerns of Stow citizens, only aerospace flight technology and aircraft systems management science majors would be allowed to take flight courses. The plan was to reduce the number of students in the program to address capacity issues and ensure quality. In addition, the university would raise admission standards. Schwartz said the suspension of admissions had not been based on financial concerns but "to ensure better flight education." The decision to keep the program, he said, was based on its merits and the feasibility of retaining it by making some cuts. Biggin, speaking for the flight students, was happy with the decision and hopeful the program would be appropriately funded and made more stringent. Moore said he had not made a recommendation to Schwartz but had outlined three possible options for him: phase out the program and lease the airport, contract with an FBO to do flight training, or keep the program. After the decision was made, the *Daily Kent Stater* suggested that the determining factor might well have been the obligations to the federal and state agencies due to their grant funding.[44]

Before announcing his decision on the flight program, Schwartz stated on March 20 he was resigning effective December 31 and planned to return to the faculty in the sociology department. The decision to suspend admissions to the flight program with the likely intention of ending it and getting rid of Paton Field had created a furor in many quarters, lasting nearly six months. Once the issue was settled in favor of the flight program and the KSU Airport, the question was whether there would be a positive rebound. The freshman class for the flight program for fall 1990 was down 40 percent, probably because the decision to reverse the suspension of admissions was not made until late May. The reduction in enrollment translated into smaller budgets for both the program and the airport.[45]

Several incidents involving KSU training airplanes in October 1990 frustrated Grossman. He reported problems with deer and one taxiing and five landing incidents: "But the real problem here is our safety record over the last two years. To put it simply, it stinks." Safety was a critical priority of the program, and no one wanted a repeat of the challenges faced over the past year. The last of the incidents triggered a suspension in flight training to allow for a full review. Additional safety measures were put into place, and the FAA and National Transportation Safety Board were asked to review the situation.[46]

Students Break Barriers

The period of the threatened flight program shutdown during the Schwartz presidency engendered an extremely high level of engagement and professionalism by the flight students to organize and effect a solution. Over its entire history, the program has produced graduates who have been a credit to Kent State. Some, through their determination and effort, achieved what none had done before. Two women who were part of the student group that fought to save the program and the airport during the turbulent period of the late 1980s went on to break barriers. With their distinguished aviation careers, they serve as models for future pilots.

Sarah Deal Burrow ('92) became the nation's first female US Marine pilot. At Kent State, Burrow worked the flight line at the airport for five years, was a member of the Precision Flight Team, and worked as a flight instructor. She joined the marines in 1991. Upon graduation from Kent State, she was commissioned as a second lieutenant and continued her training at the

Left: Lt. Col. Sarah Deal Burrow ('92), the nation's first female Marine pilot, 2017 (Courtesy of the United States Marine Corps)

Right: Capt. Stephanie Malby Johnson ('91), Delta Air Lines' first female African American captain, 2017 (Collection of Stephanie Malby Johnson)

Basic School (TBS) and then at the Air Traffic Control (ATC) School. When news came that the prohibition on women in combat roles in the US military had been lifted, Burrow became the first woman selected by the marines to attend flight school. In 1995, she was promoted to first lieutenant and earned her "wings of gold" as a naval aviator, becoming the marines' first female pilot. The Ohio Women's Hall of Fame inducted her in 1999. A 2002 deployment took her to the Middle East to fly support for Operation Enduring Freedom. Following twelve years on active duty, Lieutenant Colonel Deal retired and joined the Marine Corps Reserves. In 2008, she had a seven-month deployment to Afghanistan with the CH-53E reserve helicopter unit, flying over three hundred combat hours. Sarah Deal Burrow's advice to those who would follow her: "Be strong. Do your job. Don't expect anything special. And . . . know who your friends are."[47]

Stephanie Malby Johnson ('91) followed her dream to learn to fly and is now Delta's first female African American captain. At Kent State, Johnson had joined fellow flight students to protest the shutdown of the flight program in 1989. Following an internship with United Airlines and then graduation, she continued as a flight instructor at Kent State as well as at Freedom Field in Medina. In 1993, she worked as an airport operations agent at Burke Lakefront Airport and, after earning her airline transport pilot certificate, was hired as an airline pilot with Mesa Airlines in 1995. In 1997, Johnson joined Northwest Airlines as their first female African American pilot. She served as first officer on the 727 and Airbus 320. After Delta Air Lines merged with Northwest Airlines in 2008, Johnson went on to become a first officer for Delta on the 757/767. In September 2016, she became Delta's first female African American captain. Captain Johnson is a member of the Organization of Black Aerospace Professionals, and she has contributed her time and efforts to interest youth in aviation. She has played a role in the Aviation Career Education Academy, designed to introduce diverse students to opportunities in aviation and aerospace, in both Cleveland and Detroit. Kent State honored Capt. Stephanie Johnson with a Distinguished Alumni Award in 2011.[48]

Summary

During the eight years Michael Schwartz was KSU president, Paton Field benefited at long last from major capital improvements made possible by the 1977 FAA master plan. This would be the last major infusion of funding for the airport for nearly thirty years. The assurances made when accepting this FAA funding prevented the airport's demise from the challenge during the Schwartz era. With a new president, however, would come a new challenge.

Chapter 12

A New President, New Challenges, 1991–2006

The budget strains that KSU president Michael Schwartz had struggled with were still present when Carol A. Cartwright became the tenth president of Kent State University in 1991. Her message to the university community was that a multi-million-dollar shortfall was expected and that the current level of spending was unsustainable.[1] The KSU Airport was already under pressure to increase revenue and reduce expenditures.

First Steps under a New President

KSU Airport manager Eugene Ripple prepared an extensive report for the new president on Paton Field and the flight program, both of which, he said, had found themselves in a state of "purgatory" within academia at Kent State. His report provided background on the history, organization and administration, budget, grants, and relationships with other entities. Ripple then outlined the full range of problem areas and proposed solutions. In addition to staffing, budgets, and airplane fleet, he did not mince words on the inadequate facilities for students and the public, which were "old" or "temporary" and, most critically, lacked sufficient space. To deal with revenue issues, Ripple advocated that the airport be permitted to expand its business outreach to include land leases and additional T-hangars.[2]

However, the key issue for Ripple dealt with organization, specifically that the airport and flight program reported to different areas within the university—the airport to the business side and the flight program to the academic side. Ripple maintained that the intrinsic differences in goals and responsibilities of each impeded decision-making to the detriment of all parties, particularly the students. He proposed that Kent State create a department of aviation, bringing the two operations together under the vice president for business affairs. This, Ripple believed, would provide the KSU Airport and flight program a unified purpose for planning and a foundation for decision-making. He recommended that a committee representative of all parties be formed to discuss the issues and make recommendations to the president.[3]

The reporting lines for the airport and flight program did not change, but Ripple's report seemingly did influence President Cartwright to invest in airport improvements to address its annual shortfall of $83,000 and possibly even generate a profit. To that end, in the summer of 1992, the runway was extended by fifty feet, bringing it to a total length of four thousand feet, to make it more attractive to business aircraft. The old 80-octane fuel tanks were replaced with those dispensing 100-octane fuel to serve more aircraft and, they hoped, increase fuel sales. Also planned was the addition of a fourteen-unit T-hangar building that promised to be revenue-generating, given a waiting list dating back to 1974 for the existing T-hangar complex.[4]

Ripple retired effective June 1, 1992, having started at Kent State in the early days of flight training as an instructor in 1967. He became airport manager in 1973 and over the next eighteen years also served as coordinator of the flight program for five years. Ripple's many contributions included his hands-on role in the development and implementation of the university's first Federal Aviation Administration (FAA) master plan, which provided long-awaited capital improvements for the airport, and his support of the Organization for the Continuance of Aviation Education (OCAE) effort to save the flight program.[5]

Ripple was replaced as airport manager by another military veteran, Col. Lynn C. Feterle. Feterle had enrolled at Kent State in 1964 and joined the Air Force ROTC program, where he learned to fly. After graduation, Feterle flew C-130 transport aircraft during the Vietnam War. In 1988, he came back to Kent State for a four-year stint as the commander of the Air Force ROTC program, as well as a member of the aerospace faculty. Upon retirement from the air force, Feterle was appointed airport manager.[6]

On the morning of August 10, 1993, it fell to Feterle to report that a KSU airplane, a 1984 Cessna 172, was missing. After returning from a flight the night before, a student had locked the airplane, tied it down, and turned the key in to the dispatcher. The airplane was discovered missing at 6:30 the next morning, and after a two-hour search, KSU police and the FAA were

Left: Col. Lynn C. Feterle (in 1993) served as commander of the KSU Air Force ROTC unit and then became airport director in 1992. (Photo by Jean Angelo. Courtesy of the *Daily Kent Stater*)

Right: Flight students Jay Katz and David Vidovich pose in front of Kent 28 after returning from a cross-country trip, 1988. The airplane would be stolen from the airport in 1993. (Scott and Heidi Koeppl Collection)

notified. The airplane, with tail number N97951, was valued at $48,860. When stolen, it had enough fuel on board for one to two hours of flying time. Given that Cessna was no longer manufacturing that aircraft model, one theory was that the airplane might have been stolen for parts. The last information on the airplane was received by the KSU Police Department in April 1994. The FAA flight supervisor in St. Petersburg/Clearwater, Florida, had received a call requesting a weather briefing. The caller identified himself with the "N" number of the stolen airplane. The airplane would have been within a ten- to fifteen-mile radius of St. Petersburg. No more was heard, and the case was never solved.[7]

Frustration by flight students was brewing in spring 1994. They perceived lack of university support for the flight program and heard rumors of its possible elimination. Through OCAE, the students maintained that Kent State was deliberately undercutting the flight program and Paton Field. David Moll, OCAE secretary, characterized the attacks as "a low-key, systematic, coordinated and sustained effort to severely reduce flight training." Lawrence Kelley, vice president for business and finance, said he was not aware of any efforts to eliminate the program. Since 1991, the program was reported to have lost more than $600,000 and Paton Field almost $100,000. Moll maintained that the enrollment cap placed on the program in 1991, which forced a reduction in the number of students from 428 to 240, contributed to the revenue shortfall. OCAE's recommendations to save the program included reopening flight courses to nonmajors and creating an aerospace technology minor.[8]

Concern over the future of the flight program and the airport prompted the aeronautics program to develop a flight operations/airport master plan in March 1994. Enrollment had shrunk dramatically, in large part due to the temporary cessation of admissions in 1989. Reduced enrollment resulted in fewer flight hours, adversely affecting revenue. The master plan called for a Center for Aviation Studies and was submitted to Terry Kuhn, associate vice president for academic and student affairs. Similar to the 1991 Ripple report, the plan recommended a new organizational structure that would consolidate the aerospace flight technology program with airport operations to improve decision-making and provide more efficiency. It also proposed making the funding structure of the flight program similar to the practice at other institutions, ending the requirement that flight fees be used to cover all academic expenses as well as aircraft operating costs.[9]

Task Force Review of Flight Program and Airport Operations

In June 1994, Feterle was abruptly notified that the position of airport manager had been eliminated as a cost-cutting measure and that he would be on special assignment until his contract ended in September. On that same day, the university administration announced its commitment to the flight program and, echoing Ripple's earlier recommendation, was creating a task force to review both the program and airport operations. Cartwright charged the Flight Training Task Force to "determine the feasibility of plans for phased-in growth for the Flight program, to evaluate possible partnerships within the region, and to assess the resources required to support any recommended expansion." The task force was to submit its report by late October 1994.[10]

The task force delivered the final report in February 1995. Its central conclusion was that "the Flight Training program, in its symbiotic relationship with the Kent State airport, has the potential to grow as an asset to the University." As charged, the task force explored possible area collaborations, but it found nearby airports were either uninterested or cost prohibitive. In the end, they decided that "KSU at KSU" offered benefits for marketing, visibility, program uniqueness, and convenience for students and staff.[11]

The decline in enrollment in flight courses from 408 students in 1989 to 190 students in 1994 resulted in a 57 percent decrease in flight time, which seriously impacted revenue. The task force recommended controlled enrollment for flight training, allowing it to grow by 5 percent annually through 2001 to 264 students. It also placed "an ultimate maximum enrollment limit" of 150 students per semester until additional airport facilities were constructed. Current Paton Field facilities—the terminal, the two trailers used for flight training, and the FAA examination center trailer—were deemed to be marginally adequate. Given that increased enrollment would place more demands on these limited facilities, the task force recommended construction of a new 13,000-square-foot Aviation and Operations Center by 2001.[12]

President Cartwright accepted the report of the Flight Training Task Force in January 1996 and stated that the findings were consistent with the challenge from the Ohio Board of Regents to provide technical education and serve the needs of regional business and industry. In presenting the task force recommendations to the board of trustees, KSU provost Myron Henry maintained that "the management of the airport is the key ingredient to this plan." He also emphasized that the plan took into consideration the needs of the Stow community. In fall 1996, flight classes were opened to all students for the first time since the 1980s. By 1999, a customized flight education program was offered through the College of Continuing Studies, opening flight training to any student, staff, or community member.[13]

Land-Use Changes

The board negotiated several leases, including a seven-acre parcel of land located at the southeast corner of the airport property, an agricultural land lease for twenty acres on the east side of Paton Field, and a golf-driving range lease with Pambi Development/W. G. Lockhart Construction Company. The leases reduced the costs of maintenance and generated revenue.[14]

At the same time, Cartwright initiated the development of a long-range Kent campus master plan to envision how the university might develop and expand its land use.[15] While the focus of the 1996 final plan was on the main Kent campus, there was consideration of the airport property.

Land development surrounding Paton Field had picked up as predicted in 1977, when the state lifted its ban on sewer tap-ins. Stow's population in 1990 grew to 27,700, a 34 percent increase over 1970. Now housing developments around and near the airport had replaced the open areas. Plans called for Pambi Farms, an upscale housing development on the north side of North River Road adjacent to the KSU Airport and condominiums south of the road and Paton Field. With this development, the only remaining open area by the airport was to the east, but that became a shopping center in 1994 with the construction of the Stow Community Center.[16]

The Kent campus master plan acknowledged this rapidly developing area around the airport. It found that possible commercial or residential development of Paton Field's residual property was hampered by lack of roadway access and its orientation to the airport. One plan showed potential development with the existing airport and another showed possible total reuse of the property if there were no airport. The plan suggested that "if, at some future date, in cooperation with the FAA, the University either relocates or ceases to operate the airport, the site could be effectively redeveloped for an integrated mix of commercial, residential, and recreational purposes." This assessment foreshadowed action Cartwright's administration would take several years later.[17]

Airport Improvements

Following the flight student discontent in 1994 and the subsequent task force recommendations in 1995, the KSU Airport made modest improvements. Furniture in the terminal conference room/pilot lounge was upgraded. The FAA exam center was moved from Paton Field to Academic Testing Services on campus, and the then vacated FAA trailer was transferred to Flight Operations. That summer, Richard Schwabe lost his medical certificate and was no longer able to provide the final flight tests for students. Tom Grossman, chief pilot, became an Airman Certification Representative (ACR), granting Kent State examining authority for the first time.[18]

The FAA assessed the condition of the airport's runway in fall 1997 and found it in need of repair. An FAA grant had financed repaving in 1982. The Ohio Department of Transportation (ODOT) grants covered resealing in

Flight instructors who flew the formation flight over Stow's July 4 parade in 1997. *Left to right:* John Wheeler, Kris Palcho, Tom Grossman, Sarah (née Davis) Nicholson, Kevin Garlock, Becky (née Duke) Bauer, Richard Schwabe, Craig Barnes, Tim Palcho, and Seth Kornblum. (Becky Bauer Collection)

1989 and crack repair in 1996. In 1999, the board of trustees approved accepting an FAA grant of $710,115, which included a local match of 10 percent. The funds provided for rehabilitating the main runway, the parallel taxiway, and the connecting taxiways and terminal apron.[19]

Ruth Sitler (Kent State University Airport)

The last time Kent State had purchased new airplanes was in 1984. In 2000, the university finally acquired five new airplanes. Each of the Cessna 172s was equipped with a Global Positioning Satellite (GPS) system, a single-axis autopilot, and fully lighted instrument panel. In March 2001, the university announced the hiring of three additional faculty for the flight program, the purchase of a turbo-engine flight simulator, and a new partnership with Continental Express Airlines. The Continental Express partnership created an ab initio training program (a training partnership between an airline and flight school that establishes a curriculum that takes a pilot from early training to new airline hire) open to qualifying sophomores. It included a summer internship at the carrier's Houston training facility and, upon graduation, the guarantee of an interview with Continental Express for available pilot positions.[20]

The hire of the new faculty coincided with the retirement of Ruth Sitler, an advocate for women in aviation. Sitler had joined the KSU Flying Club in 1957 and through it earned her pilot license. She became a flight instructor at Kent State in 1987 and went on to become chief flight instructor and then flight operations manager. The National Association of Flight Instructors named her a master flight instructor in 1990. Sitler was a member in the Ninety-Nines, an international organization of female pilots, and she flew in two of the group's sponsored air races. One was the Powder Puff Derby in 1969 in which she flew with Marsha Orton in the club's Cessna 172. The pair finished third in the small airplane category, piloting the smallest airplane in the competition. Sitler would speak of this achievement as one of the highlights of her career. Her doctoral work at Kent State culminated with a dissertation that explored gender differences in learning to fly. Throughout her career, she served as a mentor to many and a role model for women. Sitler was honored by being the first to receive the Van Deusen Distinguished Professor award.[21]

Airport Response to 9/11

Reaction to the terrorist attacks of September 11, 2001, were immediately visible at Paton Field. The national flight ban called for the airport to lock all airplanes, lock the hangar doors, and turn off the fuel pump. Following the lifting of the ban, Isaac R. Nettey, senior director of aeronautics, explained that all flights for the next week would be local and none would be solo to avoid anything approaching "sympathy action," or copycat crime.[22]

FAA Master Plan

The first step toward Cartwright's move to close Paton Field and relocate the flight-training program began late in 2001. The university accepted an FAA planning grant of $150,000 to update the 1974 airport master plan and the 1984 airport layout plan, thus beginning a new airport master plan process. The FAA master planning process assesses current conditions, forecasts demand, and develops alternatives to meet goals and objectives over a twenty-year planning horizon. This provides the basis for the FAA to award grants for modernizing existing airports. Richland Engineering Ltd. of Mansfield, Ohio, was selected as the consultant to lead the KSU project.[23]

As the master plan process was getting under way, a report surfaced in August 2002 that the university had discussed closing Paton Field with the FAA and moving the flight program to the Portage County Airport. US representative Steven LaTourette, a member of the House Transportation Committee, was commissioned by Kent State to serve as liaison between the university and the FAA. LaTourette said the rationale for the proposal was lack of space for airport expansion and safety concerns of nearby residents. He admitted the FAA was not receptive to the proposal "because the FAA encourages airports, not the closing of airports."[24]

Albert Beckwith, although a member of the master plan standing committee, was unaware of these discussions and the involvement of Congressman LaTourette. Concerned after seeing the report of the meeting in the newspapers, Beckwith created a petition to garner support for the airport on Aviation Day in September. He acknowledged the university was free to move its flight program to the Portage County Airport, but he maintained that Paton Field should then revert to the City of Stow as a general public-use airport. He began making contacts immediately, reaching out to all stakeholders and government officials. An ODOT official underscored that Kent State was obligated under grant assurances to operate the airport in a safe manner until 2019. Beckwith advised Stow City Council in October 2002 that his petition in support of the airport had 376 signatures. He also informed the council that in the first seven months of the year, 111 individuals had flown into the airport and rented cars or vans from his business, demonstrating Paton Field's value to the community.[25]

In July 2003, the FAA met with Kent State's master plan committee and said the plan needed more work. FAA officials indicated disappointment with progress to date and that a plan was not now expected until 2004.[26] The first of several public meetings on the airport master plan was held on December 10, 2003. The plan forecasted an increase in general aviation operations of 40 percent over the next twenty years, including use by larger aircraft. The issue was how to best accommodate this expected growth and support the future needs of the flight program.[27]

A standing-room-only crowd at the January 2004 public meeting had its first look at a number of alternative proposals. The most vocal comments were for keeping Paton Field as it was or moving its operations to the Portage County Airport in Ravenna, Ohio. A group of residents in opposition to the airport formed Communities Against Flight Expansion (CAFE). At the open meeting in February, the university offered three alternatives. Plan 1a would keep Paton Field as it was. Plan 1b provided a modification that called for widening the runway by fifteen feet. The third alternative, called Plan 7, would relocate operations to the Portage County Airport and decommission Paton Field. Attendees were assured that public input would be taken into account. Following a final decision, the plan would be taken to the board of trustees and then forwarded to the FAA.[28]

Supporters formed a Friends of the Airport group, highlighting the airport's potential for contributing to economic development. It supported widening the runway to enhance safety, installing an automated weather beacon, and improving leasing arrangements for businesses. The Stow-Munroe Falls Chamber of Commerce passed a resolution calling for retention of Paton Field and for future development surrounding the site. CAFE presented Stow City Council with 300 signatures of those opposed to any expansion of the existing airport. The group also sought an immediate ban on all jet aircraft at the airport. The residents were concerned that even widening the runway would increase traffic and noise from jets and larger turboprop commuter airplanes.[29]

In February, KSU provost Paul L. Gaston sent a memo to David Creamer, vice president for business and finance, and Shirley Barton, executive dean of the regional campuses, seeking more information about the aeronautics program preference for the status quo. The faculty of the aeronautics program replied to Gaston's questions in March. Laying out their reasons for staying in Stow, they cautioned that there was a lack of "a clear institutional understanding of the magnitude of the financial, bureaucratic, and legal or political costs that would be required to decommission and subsequently sell off Kent State Airport." The faculty's recommendation was that direct discussion take place between KSU leadership and the FAA.[30]

Provost Gaston was not convinced by the faculty's arguments to remain in Stow. At the faculty senate meeting on April 12, Cartwright announced that Gaston supported closing Paton Field and moving the flight-training program to a nearby airport. After further review, Cartwright said she would forward the recommendation to the board of trustees for their endorsement. Prior to the upcoming board meeting, Irene Porter, the FAA official charged with overseeing Ohio airports, wrote Cartwright advising her that it was unlikely that the FAA would go along with closing the airport in Stow, stating that "the vital role this airport plays in the national transportation system . . . is inconsistent with a request for closure." Because Kent State had purchased land with the federal grants, the assurances that it had agreed to would not expire. Porter also noted that if the FAA consented to the closing, the university would be forced to pay back the $3.9 million it had accepted in federal grants to free itself from ownership.[31]

Decision to Decommission the Airport

On May 25, 2004, the board of trustees voted in support of decommissioning Paton Field and transferring air operations to a nearby airport. The board's decision was forwarded to the FAA. The FAA decision would be final and might not be forthcoming for five to ten years. Creamer noted that by accepting the FAA grant in 1998 for over $734,000 for the repaving project, Kent State was obligated to keep Paton Field open until 2018. He felt there was a compelling case that the grant money should follow the flight program if it relocated. Also, if the university then sold the current airport site of 287 acres, it would have an estimated $30 million to apply to the flight program in its new location. The preferred alternative airport was the Portage County Airport, but they would also consider other sites. The move to Portage County would cost approximately $5.9 million.[32]

Following the board's action, Cartwright responded to Porter that the university was steadfast in its decision to close Paton Field despite Porter's comments that it was highly unlikely the FAA would approve closure. Cartwright acknowledged that seeking release from federal debt obligations was "indeed serious business." Munroe Falls City Council stated its opposition to airport expansion or growth in flight operations. At a Stow City Council meeting, CAFE representatives vehemently expressed their concern over airport expansion and the detrimental effect it would have on property values. They took Porter's letter as an indication that expansion was inevitable and urged the council to pass a resolution in opposition. The council took no action, given that the decision was in the hands of the FAA. Ron Alexander, an at-large council member, was quoted as saying, "We came to the airport, the airport did not come to us."[33]

Beckwith, along with the Friends of the Airport, continued to make a case for keeping the airport in Stow. The group circulated a petition on Aviation Day in fall 2004, seeking support and urging residents to vote against politicians who did not support the airport being in Stow, singling out Congressman LaTourette.[34]

Annual Air Show Resumes

In fact, it was the annual air show for the campus and community that provided tremendous positive attention for the KSU Airport during the Cartwright years. Following a twelve-year hiatus, the event was brought back in 1995. Formerly Air Expo, it was now called Aviation Day and proved to be an exceedingly popular event, providing an opportunity to generate goodwill with the community. Aviation Day '95 drew 1,800 to 2,000 people. As would be typical for many years to come, the event featured a pancake breakfast served by the Stow-Munroe Falls Kiwanis Club, hot air balloons, and airplane rides. One flight instructor called the event "a handshake with the community and a chance to give some people a look at aviation."[35] In the years following, the event featured vintage airplanes as well as the popular flybys from military aircraft such as the F-15.[36]

The airport planning committee reached out to the City of Stow, leading to incorporation of its annual "A Taste of Stow" into the aviation event in 2001. It was then called "Aviation Day with Flair, a Taste of Stow," with the addition of culinary treats, an artist village, multicultural performances, and children's activities. In 2001, attendance grew to 20,000. Dennis Baden, who served as chair of the planning committee for many years, recalled that people came early for breakfast and "many stayed much of the day with the crowds swelling right before the F-15 flybys."[37]

The time and cost of hosting the event rose along with its growth. Attendees were charged in 2005, causing uneasiness among organizers and still failing to cover expenses. The City of Stow also had concerns about the cost of the event. All of these factors, added to Kent State's interest in closing

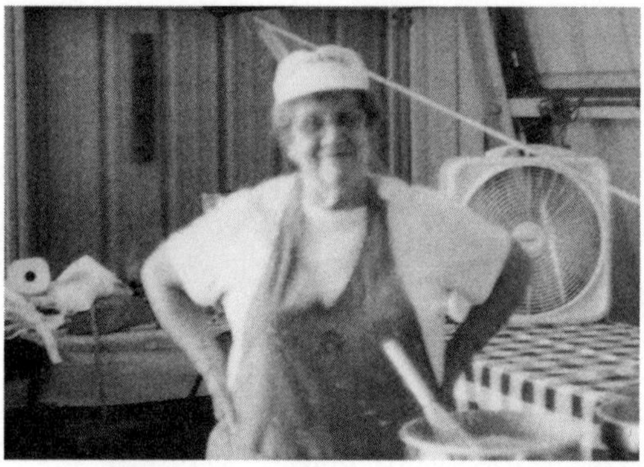

Top left: Aviation Day with Flair, a Taste of Stow, Kiwanis pancake breakfast, Grace Jensen, 2001 (Collection of the Kiwanis Club of Stow-Munroe Falls)

Top right: Aviation Day with Flair, a Taste of Stow, Tom Stephan, announcer, 2002 (Collection of the Kiwanis Club of Stow-Munroe Falls)

Above left: Aviation Day with Flair, a Taste of Stow pin, 2004 (William W. Schloman Jr. Collection)

Above center: Aviation Day '99, F-15 Eagle flyby (Photo by William D. Schloman)

Above right: Aviation Day with Flair, a Taste of Stow poster, 2004 (Kent State University Airport)

Right: Aviation Day 2000, Beechcraft 1900D operated by Continental Express out of Cleveland Hopkins International Airport (Photo by William D. Schloman)

Paton Field and moving the flight program, led the airport to end its sponsorship after the 2005 event.[38]

Efforts to save the event were successful when the Stow-Munroe Falls Kiwanis Club stepped in and worked with Isaac Nettey, aeronautics program director, to ensure the event could continue. The Kiwanis Club had a longstanding commitment to the event since 1996, when they first provided the pancake breakfast. In 2006, the club planned to fund the event through corporate sponsorships, alumni support, and voluntary contributions. They sent a letter to a hundred local businesses asking for donations. The event's name was changed to Aviation Heritage Fair, and it continued under that name through 2015.[39]

Two groups remained active over the fifteen-year period of the Cartwright presidency. The KSU Flying Club continued its long history of offering members the opportunity to rent airplanes at competitive rates, as well as providing social and aviation-oriented activities. The KSU Precision Flight Team competed six different years in the National Intercollegiate Flying Association (NIFA) Region III meet and hosted the event in 2004. In 2005, the team placed second at the regional meet, qualifying to advance to the national competition for the first time in nineteen years.[40]

Despite the turmoil and uncertainty surrounding Paton Field's future, 2006 brought some good news. In January, two Cessna 172SPs were delivered, the first new airplane acquisitions since 2000. Later, a Piper PA-44 Seminole was acquired and named in honor of Cartwright. In April, the flight program was accredited by the Council on Aviation Accreditation (now the Aviation Accreditation Board International) for having met the criteria for curriculum, faculty, and administration. Kent State was then the first and only accredited flight program in Ohio.[41] This accreditation was a major factor in the School of Technology becoming the College of Technology on July 1, 2006. Accreditation was successfully renewed in 2016.[42]

Summary

Carol Cartwright announced her retirement after fifteen years as president, effective June 30, 2006.[43] She dealt with serious financial problems at the university throughout her tenure. Cartwright's presidency featured her firm support for decommissioning the airport and moving the flight program from Stow as recommended in the 2004 FAA master plan. Whereas Michael Schwartz hoped to relieve Kent State of both Paton Field and the flight program, Cartwright's plans were more nuanced. She, too, hoped for an end to the KSU Airport in Stow but, by all accounts, was not pushing for an end to the flight program or discounting the need to reestablish necessary airport support elsewhere. It would take another ten years and the administrations of the next two university presidents to bring resolution to Paton Field's future.

Chapter 13

Reaching the Century Mark

The two Kent State University presidents who followed Carol Cartwright were able to secure Paton Field's future. Lester A. Lefton served as university president from 2006 until 2014, followed by Beverly J. Warren. Lefton inherited the board of trustees' decision to decommission the Kent State University Airport, but there had been no public mention of any follow-up steps being taken since that 2004 action. Behind the scenes between 2004 and 2006, Kent State had updated the master plan to indicate the flight program was to move to Portage County.[1]

Early in 2007, Albert Beckwith learned that the new administration was reviewing the master plan recommendations and had called for a meeting with Federal Aviation Administration (FAA) officials and representatives from the Portage County Regional Airport Authority. The authority was in the process of updating its FAA master plan. To address its airport's financial difficulties, there was interest in expanding the runway to five thousand feet to attract corporate jets and stimulate economic growth. Even more benefits could be realized if Kent State relocated its flight program to the airport. Concerned Portage County citizens voiced opposition to a possible merger with Kent State, maintaining it would bring noise and cause safety concerns, and they questioned whether there were any economic benefits. Lack of progress in talks with Kent State that spring prompted airport officials to share with the FAA their frustration, only to be reminded that the decision to relocate was Kent State's to make.[2]

Beckwith also expressed frustration to FAA officials about Kent State. He believed that, to justify Paton Field's existence, the university was working to convince the FAA and the public that there was insufficient traffic beyond that of the flight school. He also believed the university was trying to shut him down by not promoting the airport and not providing maintenance to the public's or his airplanes. Beckwith also criticized Kent State's transfer of an estimated $22,000 in FAA grant money to the Portage County Airport for its use. Beckwith's concerns about the university's intentions prompted him to contact US representative Steven LaTourette for any update he could provide. Congressman LaTourette's office responded it had no current information and reiterated that to close the airport and relocate it to Portage County, Kent State would have to repay all the federal airport improvement funding that it had received.[3]

Despite the uncertainty over the KSU Airport's future, the university submitted an Airport Layout Plan (ALP), as required by the FAA, in January 2008. Rather than depicting future development, the ALP provided an "as is" map of the airport, which the agency conditionally approved in April. The approval reiterated that the FAA did not accept the recommendation in the 2004 master plan to decommission the airport.[4]

From 2009 through 2010, there were conflicting opinions within the university as to the course of action for Paton Field. Gregg Floyd, who had joined the university as vice president for finance and administration in September 2008, directed Anne Brown, associate vice president for business and administration services and the administrator with responsibility for the airport, to review the situation and make a recommendation. Discussions were held with all parties, including the FAA. Brown's recommendation was that a new master plan was needed to provide a fresh look.[5]

A New FAA Master Plan

In 2012, the FAA provided a $600,000 planning grant for a new master plan that would be completed in 2016. Through the planning process, Kent State could reconsider the earlier decommissioning recommendation that the FAA found to be in conflict with earlier assurances. The ALP would be updated, establishing eligibility for the funding needed to improve and maintain the dated facilities. And the plan would use new FAA airport design standards and documents.[6]

C&S Engineers, Inc., of Cleveland was chosen as the primary contractor for the master plan. A steering team chaired by Brown and a community liaison group directed the effort. Beckwith was a member of the community group and presented his own master plan improvements. Regrettably, Beckwith died in October 2014 and did not see the successful adoption of a plan that would keep the airport in Stow.[7]

During the master plan process, the KSU Airport was contacted by the FAA in 2013 and invited to apply for a sustainability grant. The FAA awarded the university a $114,000 grant—one of three that the FAA awarded in Ohio. This grant gave Paton Field the opportunity to serve as a model for like-sized airports. The goals of the sustainability plan were to reduce environmental impacts, achieve economic benefits, and increase integration with local communities. This planning would be integrated into the larger airport master plan.[8]

Albert Beckwith, owner of Commercial Aviation Corporation and airport advocate, speaks on September 13, 2014, to a large crowd about his long history with the KSU Airport and his hopes for its future. He would die less than three weeks later. (Photo by William D. Schloman)

When Lefton's presidency ended in 2014, numerous construction projects on campus and in the city of Kent contributed to a transformation of both the city and the university. The College of Technology (now called the College of Applied Engineering, Sustainability, and Technology) had gotten a new building on campus, but there had been no investment in Paton Field.[9]

When Beverly Warren became president, work on the FAA master plan had been under way for two years and had identified three objectives: accommodate growth in the flight-training program, comply with FAA grant assurances, and move toward financial independence. Background work for the plan included assessing Paton Field facilities, finances, and community outreach. The condition of the facilities ranged from fair to poor, specifically for the main runway, parallel taxiway, apron, T-hangar taxiways, runway, taxiway lights, drainage issues, main hangar, maintenance hangar/terminal building, trailers, T-hangars, and vehicle parking.[10]

The financial review found airport revenue had increased over the past five years but was still insufficient to cover expenses. The capital improvement account used for major maintenance and a local grant match was funded through the land lease with Pambi Development and a fuel tax reimbursement. Kent State projected that enrollment in the flight program was to increase and equated it to increased airport operations. The noise analysis done as part of the environmental assessment determined that noise abatement procedures and levels were compatible with airport operations.[11]

The community outreach assessment found the KSU Airport active on numerous fronts. These included supporting the Aviation Careers Academy in partnership with local schools, assisting with the Nikki Kukwa Memorial Aviation Camp for young women, cohosting the Aviation Heritage Fair with the Stow-Munroe Falls Kiwanis Club, providing tours to local groups, serving as a site for first responder training and certification, and collecting food and supplies for local animal shelters with the Props-4-Paws program.[12]

Following this review, the committee developed four alternatives for the KSU Airport's future. The first three alternatives kept the airport and flight program in Stow with varying levels of development. The fourth alternative kept the airport as a public facility for general aviation and relocated the flight training to a nearby airport. The committee held three public meetings over a sixteen-month period.

The preferred alternative called for maintaining the airport in Stow as a public-use airport, but with a focus on flight training. Paton Field would not be actively marketed to attract corporate aviation users. This approach would serve the university's academic mission and address community interests. The runway length would not change, and the current infrastructure would be maintained and improved, not expanded. On September 9, 2015, the board of trustees approved the preferred master plan alternative and authorized submission of the plan to the FAA for its approval. This action superseded the 2004 resolution by the board to decommission the airport.[13]

FAA Master Plan Approved

FAA approval of the new master plan came on July 27, 2016.[14] This solidified the airport remaining in Stow and provided for the first major investment since the 1977 plan. The approved plan would be implemented in three phases over the twenty-year planning period (2015–35). The initial emphasis would be on addressing safety concerns, then undertaking repairs, and finally making enhancements.

Phase 1 (2015–19) includes runway rehabilitation, taxiway and apron improvements, an Automated Weather Observation Station (AWOS), obstruction mitigation, storm water management, and electrical system upgrade. An AWOS had been a hoped-for acquisition for some time because it provides valuable, real-time local weather information for pilots flying in and out of an airport. The addition of the AWOS at the KSU Airport could bring a change to its well-established 1G3 airport identifier.[15]

Phase 2 (2020–24) gives high priority to a new terminal/academic building.[16] As early as 1960, when the university had ownership of both the

Above left: KSU Airport terminal lobby, 2017 (Photo by William D. Schloman)

Above right: Instructor and student with flight simulator in trailer at the KSU Airport, 2017 (Photo by William D. Schloman)

Left: Personnel performing routine airplane inspections in the maintenance hangar, 2017 (Photo by William D. Schloman)

airport and its buildings, the board of trustees made a failed capital request to the state for a new aviation education building at the airport. Many requests followed over the years. Now, with the master plan recommendation and full university support, plans for a building began to materialize in 2016.

Progress toward an Airport Education Building

The planned 17,800-square-foot facility with "collaborative spaces, upgraded virtual reality flight simulators, faculty and administrative offices and pilot and passenger support spaces" is slated to open in fall 2019. The new building will house flight training and operations, but not maintenance. One-quarter of the space will be for airport operations and the flying public. Kent

Architectural rendering of proposed new aeronautics education building, 2017 (Kent State University)

State will provide the services typically offered by a contracted fixed-base operator, the last being Beckwith's Commercial Aviation Corporation, which went out of business in 2016. The trailers used in support of flight training are to be removed to allow for an expanded and improved parking lot.[17]

The new building, the FedEx Aeronautics Academic Center, is financed through major gifts, including $6.5 million in donations and pledges from FedEx. Construction of the building was approved by the board of trustees on March 7, 2018.[18]

Phase 3 (2025–35) of the master plan calls for rehabilitation of the main or community hangar and construction of a new maintenance hangar. The main hangar, dating from 1945 to 1946, is the iconic feature of the airport, recalling an earlier time in aviation. Its preservation will sustain that connection to Paton Field's early history and maintain it as a symbol for the airport. This phase also calls for four new fourteen-unit T-hangars. Two will be used for KSU airplanes, one as a replacement, and the fourth for expansion.[19]

Tradition of the Air Show

The FAA master plan notes that the annual air show plays a role in community outreach. Always a popular event for the campus and community, it promotes aviation and the university's aeronautics program and engages residents with Paton Field. Faith Cook, Stow-Munroe Falls Kiwanis Club cochair in 2009, called the fair "a true community event."[20] In 2016, the name changed from the Aviation Heritage Fair to the Aeronautics Fair.

As had been the case for the past one hundred years, residents still could take an airplane ride at this airport, now offered by the Kent State University Precision Flight Team. Vintage airplanes on display, such as the North American B-25 Mitchell and Douglas C-47, were always crowd-pleasers. Attendees were treated to an appearance by Goodyear's new airship *Wingfoot One* in 2014 and then again in 2017, in both years piloted by KSU graduates. There were often hot air balloons and for a number of years a flyby performed

Above: Goodyear's *Spirit of Goodyear* blimp departs after unhooking from its temporary mooring mast, September 12, 2009. (Photo by William D. Schloman)

Left: Goodyear's first Zeppelin NT, *Wingfoot One,* at the September 18, 2015, Aviation Heritage Fair, piloted by three KSU alumni: Michael Dougherty, Adam Basaran, and James Kosmos (Photo by William D. Schloman)

Aviation activities for kids. Billy Schloman is pushed by his younger brother Jack in a pedal airplane, 2015. (Photo by William D. Schloman)

Georgie's Gal, a B-25J Mitchell from the Liberty Aviation Museum in Port Clinton, Ohio, arrives at the September 14, 2013, Aviation Heritage Fair. (Photo by William D. Schloman)

by an F-15 or F-16. The hangar was filled with exhibits by the flight program, the MAPS Air Museum, model airplane club, and local businesses. Activities were provided for children, along with the opportunity to fly a simulator. In 2013, World War II veterans were added to the program as speakers.[21]

Starting the day at the air show with the pancake breakfast is a tradition for many. In 2017, student groups provided the breakfast, marking the end of the Stow-Munroe Falls Kiwanis Club's twenty-year run. The club's participation had extended to giving a $1,000 scholarship to a flight student from its pancake breakfast profits.[22]

Success of the KSU Precision Flight Team

The KSU Precision Flight Team enjoyed considerable success during the Lefton years. The team qualified to participate in the National Intercollegiate Flying Association (NIFA) National Safety and Flight Evaluation Conference (SAFECON) competition several times. In 2008, the team got first place in the regional competition, which was hosted at Kent State. Associate dean Isaac Nettey commented this was the first top win at regionals "since any of the students were born." The ultimate achievement occurred in 2010, when the team won the Loening Trophy at the national competition. This NIFA award "is a perpetual trophy presented annually to the outstanding all-around collegiate aviation program in the nation. The Loening Trophy is the rarest and oldest of all collegiate aviation awards." The Ohio House of Representatives recognized the team's achievement at the statehouse on November 16, 2011.[23]

In 2007, Calvin Carstensen, a flight clinic student, offered to purchase and donate an airplane for the KSU fleet. Working with Tim Palcho, chief flight instructor, a 1968 Cessna 150H was obtained. Named "The Spirit of the Flying Deuce," the airplane was used by the KSU Precision Flight Team for its competitions.[24] In 2012, the team hosted the Region III competition for the first time since 2007.

Loening Trophy, awarded to the Kent State University Precision Flight Team in 2010 as the best all-around collegiate aviation program in the country. *Left to right:* KSU president Lester Lefton, Evan Taylor, Myles Grimm, Jeff Adelman, Chris Robison, Conrad Plasterer, John Torok, Prof. Richard Mangrum, and KSU provost Robert Frank. (Kent State University Libraries. Special Collections and Archives)

Lining the runway for the landing competition at the October 15, 2008, Safety and Flight Evaluation Conference at Kent State University are Tim Palcho, Ben Satyshur, and Jim Sebastian. (Photo by William D. Schloman)

Commemorating the Past

Richard Schwabe, who had been with the flight-training program since its inception at Kent State in 1966, died in March 2011 at age eighty-nine. For forty years, he had served as a flight instructor and FAA examiner, as well as an aerobatics performer at numerous air shows. At a reunion held in his honor in 1997, 150 former flight students returned from 45 states and 4 countries to show their gratitude to the man who was their instructor, friend, and mentor. They recalled his regular presence in the terminal lobby launching cookies at passing students and, when catching the uninitiated unaware, creating a teachable moment "to stay on your toes and always maintain your situational awareness." Schwabe left a legacy by imparting the highest standards of flying to hundreds of students. The college established a scholarship in his name in 2008. To honor his passing, a memorial flyover was made on April 1, 2011, with faculty and alumni taking every available KSU airplane aloft, led by a formation of five airplanes.[25]

In 2016, the flight program reached the milestone of fifty years of university-provided training. The history traces back to the postwar curriculum that Andrew Paton developed and his vision of Kent State training pilots, using its own airport as an educational facility. In 2006, the aeronautics program became one of six Kent State bachelor degree programs accredited

Briefing by Tim Palcho for the Richard Schwabe tribute flight, April 1, 2011 (Photo by William D. Schloman)

with Aviation Accreditation Board International. As of 2016, the program reported 198 majors, who upon graduation were prepared to be pilots with all necessary certificates and ratings.[26]

Looking toward the Future

In 2017, the board of trustees created the College of Aeronautics and Engineering, a renaming and refocusing of the previous College of Applied Engineering, Sustainability, and Technology. This certainly represented another long journey from the early days of being the Aviation Technology Division in the Department of Industrial Arts. Looking ahead at that juncture, interim dean Robert G. Sines Jr. underscored the college's commitment to bringing a new facility to the airport, upgrading the fleet to ensure the best equipment for training, and producing the finest pilots for the aviation industry.[27] In 2018, Christina Bloebaum, a noted aerospace engineer, was appointed dean of the college.

As the KSU Airport's century ends, how fitting that the Ohio Aviation Association in 2017 awarded "Airport of the Year" to Kent State University. This award acknowledges that the airport has made "extraordinary accomplishments" in moving its airport program and Ohio aviation forward. David Poluga, KSU Airport manager, received honorable mention for "Airport Manager of the Year."[28]

With the decision to keep the airport in Stow and with renewed support from the university, Poluga sees a future built on sustainability, not expansion, with a new emphasis on doing things better and smarter, and an ongoing focus on the environment and the need to build stronger community relations. He believes, "A new era of aviation is unfolding at Paton Field."[29]

Above left: "Airport of the Year" awarded to the Kent State University Airport by the Ohio Aviation Association in 2017 (Photo by William D. Schloman)

Above right: David Poluga, KSU Airport manager, in his office, 2017 (Photo by William D. Schloman)

Right: KSU Airport student employees and staff, *eInside,* June 26, 2017. *Standing, left to right:* Jason Shelbourn, Ben Luecke, Kareem Rogers, Imani Matthews, Thomas Friend, and David Poluga. *In front, left to right:* Donald Schettler and Aaron Hooper. (Courtesy of Kent State University)

Bottom: Thomas Friend, KSU Airport operations coordinator, at the "line service desk," 2017 (Photo by William D. Schloman)

Epilogue

Activity at Stow Field in its early years reflected aviation's budding start. As the airfield became established, it matured along with Stow, Ohio, and survives today as one of the city's oldest facilities. Similarly, the first decades of the Kent State University Airport were in step with those of the newly created school in neighboring Kent. When Kent Normal School was founded in 1910, Rev. Edward J. Smith had just built the large house on his model farm. When the school became Kent State College in 1929, Stow Field was a known airfield, despite being located in a rural community with a population that was less than 2 percent of Akron's.

That Stow Field achieved the prominence it did is a credit to two individuals in particular who used their standing in the area to promote the airfield. Fred Smith, although not a pilot himself, was a politician and adept at making connections to promote his enterprise. Those qualities helped establish Stow Field as Akron's de facto airport until the establishment of Akron Municipal Airport in 1928. The crowning examples of bringing attention to Stow Field were the two huge aerial exhibitions held there in the summer of 1925. The next effective promoter of the airfield was Hugh C. Robbins, who was both a well-regarded pilot and entrepreneur. Robbins was a familiar sight at air events and regularly mentioned in newspaper accounts. In addition to being the Waco distributor for northeast Ohio, he expanded the visibility of his business into Akron with Aviation College. Of all the events Robbins sponsored at Stow Field, none would match the notoriety accorded Frankie G. Renner's altitude attempt in 1931.

The story of aviation's development in this region ties Stow Field together with Akron Municipal and Mid-City Airports. Each played a role in connecting the public to aviation. While the airports competed for the same business early on, their personnel were well known to each other and often moved between the various operations for employment. For many years, Rudy Van Devere, whose start was in Akron, would be the link among all three airports. His role in providing the flight training for the Civilian Pilot Training Program (CPTP) brought KSU students to Akron Municipal and

Mid-City Airports when the university's own airport was unflyable. Mid-City continued as a key alternative to the KSU Airport until its closure in 1956.

As Paton Field approaches its centennial, it has the distinction of being the longest-surviving public-use airport in Ohio. Contributing to its longevity was the continuity of ownership that provided constancy even in challenging times. The short life of the Ohio Flying School and Transport Company demonstrated for the Smiths that the newly created airfield could be a legitimate business, prompting them to hold on to the airport property over the next years, even as operators came and went. Rudy Van Devere followed as Stow Field's owner long enough to be a bridge to KSU ownership, averting possible development. When World War II began, other small airports in the area went out of business. But despite little activity, Stow Field survived due to its close connection with Van Devere and his Akron Airways, Inc., business.

The university represented state institutional ownership, seemingly providing the best guarantee for the airport's continuing existence. But as the threats of 1989 and 2004 showed, that in and of itself was not necessarily sufficient. It took the assurances agreed to when accepting Federal Aviation Administration (FAA) funding to maintain the airport. With approval of the most recent FAA master plan and university support, Paton Field now seems truly anchored in Stow.

Three key decisions made after Kent State entered this story led to the airport that exists today. Extending over thirty-five years, those were participation in the CPTP in 1939, creation of an airport study committee in 1963, and acceptance of FAA funding in 1975.

The decision by KSU president Karl Leebrick in 1939 for Kent State to participate in the CPTP brought aviation to the university for the first time. Support by the State of Ohio for the CPTP prompted it to direct Kent State and four other state universities to each purchase an airport for that training. This led to Kent State's purchase of Stow Field in 1942 and established collegiate aviation in Ohio with four of the original five college programs still in operation today.[1] CPTP involvement had other important effects. It brought Andrew W. Paton in as an instructor and formed his vision for a KSU flight program, providing him the basis for the curriculum he developed in 1946. The CPTP experience also led to postwar participation in other federal programs that underscored the viability of a Kent State–operated flight program and provided the needed revenue to make that start.

The second key KSU decision affecting the airport was made by KSU president Robert I. White in 1963. Influenced by Paton, White created an airport study committee to determine how the property might best be used.

The committee's key recommendation was for the university to take over flight training and use the airport as an educational facility. This would bring it under total university control and eliminate the problems caused by using a contracted operator. The recommendation came to fruition in 1966 with the establishment of Kent State's flight program and a new role for its airport. Andrew Paton's vision became a reality, making it fitting indeed that the hangar and airfield bear his name.

The third critical KSU decision for the airport's future related to finances. Lacking the funds to address acute maintenance and capital improvement needs led KSU president Glenn A. Olds in 1973 to seek FAA funding after all other possibilities had been exhausted. Accepting federal funding obligated the university to honor assurances agreed to with the FAA, saving Paton Field later from threats by two different university administrations. Implementation of the long-overdue improvements recommended in this first FAA master plan began in the early 1980s. These would be the last significant changes until another FAA master plan was approved in 2016.

During the nearly eighty years of KSU ownership, numerous individuals advocated for the airport. First and foremost was Andrew Paton, who advised two KSU presidents of the airport's educational role in support of flight training. Later supporters of note were the flight students who created the Organization for the Continuance of Aviation Education (OCAE) to oppose KSU president Michael Schwartz's attempted shutdown of the flight program and airport. Their activism and well-crafted campaign engaged the broader campus and aviation community. Later, KSU president Carol A. Cartwright's actions to close the airport in Stow and move the flight program to Portage County faced determined, organized resistance led by Albert C. Beckwith, the airport's longtime fixed-base operator. These individuals are representative of other advocates who spoke for the airport over time and made a difference.

Support for women in aviation at this airfield goes back to Frankie Renner, one of the first female pilots in the Akron area and Ohio. Renner went on to earn additional pilot ratings, participate in aerial exhibitions, and become an airplane salesperson. Her failed altitude record attempt at Stow Field did not diminish her status as a trailblazer. Renner's efforts, along with those of her fellow members of the Squadron of Death, demonstrated the possibility of new pursuits for women. A half century later, the KSU Airport would again be the training ground for female pilots who would distinguish themselves by their professional accomplishments and serve, as Renner had, as new role models.

Throughout this history, individuals brought life to the airport's story. Some of those who played a key role in the life of Paton Field have been highlighted in this account, but many others also contributed. It is worth reflecting, too, on the thousands who experienced aviation on the airport's property over its hundred-year history. Some were in the crowds at the aerial exhibitions. Many others had their first airplane ride. For flight students, the memories of learning to fly at this field are indelible. This story underscores Andrew W. Paton Field's rich legacy.

Notes

Abbreviations

ABJ	*Akron (OH) Beacon Journal*
BOT	Kent State University Board of Trustees records
Bowman papers	George A. Bowman papers
Cartwright papers	Carol A. Cartwright papers
CPD	*Cleveland Plain Dealer*
DKS	*Daily Kent (OH) Stater,* as well as its related titles (*Kent Stater, Summer Kent Stater*)
K-Trib	*Kent (OH) Tribune,* http://www.kentohiohistory.org/kent_tribune.html
R-C	*Record-Courier* (Ravenna, OH), as well as its related titles (*Evening Record, Kent Courier*)
KSU Trustees	Kent State University. Board of Trustees
KSUSCA	Kent State University Libraries. Special Collections and Archives
Leebrick papers	Karl Clayton Leebrick papers
Olds papers	Glenn A. Olds papers
Schwartz papers	Michael Schwartz papers
White papers	Robert I. White papers

1. The Early Years through 1922

1. We based our determination that the Kent State University Airport is the oldest-surviving public airport in Ohio on identifying currently operating public-use airports in Ohio using Federal Aviation Administration (FAA) data. *Airport Facilities Data,* s.v. "Ohio/Public-Use," Federal Aviation Administration, Apr. 26, 2018, https://www.faa.gov/airports/airport_safety/airportdata_5010. We then checked whether any of these airports existed prior to 1920 using the following sources and supplementing with internet searches as needed: Bruce Eytinge, *Flying Guide and Log Book* (New York: John Wiley and Sons, 1921), 110–13; US Coast and Geodetic Survey, Federal Aviation Administration, and National Ocean Survey, *Sectional Aeronautical Charts: United States* (Washington, DC: US Coast and Geodetic Survey, 1932–33, https://www.loc.gov/resource/g3701pm.gct00089/?sp=38&st=gallery.

2. Summit County Fiscal Office, *Grantor-Grantee Index,* s.v. "Edwin J. Smith" (book 169, page 299, accessed June 9, 2016).

3. "Beacon Journal Woman's Page," *ABJ,* Sept. 4, 1914; "Native of Brooke County"; "Fish Creek," *Community Church News* (Stow, OH) 9, no. 19 (May 8, 1942): 6, http://www.smfpl.org/files/lh/vol09/THE_COMMUNITY_CHURCH_NEWS_VOLUME_09_1942p235.pdf. Today, the house remains a presence on Kent Road and is home to a wine bar. It is believed the barn came down in the 1930s.

4. *1910 United States Federal Census, Supplement for Ohio,* table 1: Population of Minor Civil Divisions, 590; James M. Blower, *Northern Ohio Traction Revisited* (Akron, OH: James M. Blower, [1968?]), 175–76; Craig Sanders, *Akron Railroads* (Charleston, SC: Arcadia, 2007), 107.

5. "In My Window for a Few Days Only," advertisement, *ABJ,* Nov. 5, 1909.

6. Tim Roske, "A Flight Back to Kent 1911," *DKS,* Sept. 24. 1986; Linda Joseph, "Memories of 1911 Live Anew," *R-C,* Sept. 25, 1986; "Calbraith Perry Rodgers," Barron Hilton Pioneers of Flight Gallery, Smithsonian National Air and Space Museum, ac-

cessed June 21, 2017, http://pioneersofflight.si.edu/content/calbraith-perry-rodgers0.

7. "Aviation Schools Merge," *Elyria (OH) Chronicle-Telegram,* Feb. 4, 1913.

8. "Silver Lake Airport," Village of Silver Lake, OH, accessed Oct. 27, 2015, http://www.villageofsilverlakeohio.com/airport.html; "Silver Lake Park," *CPD,* July 2, 1912.

9. "Biplane Fails to Glide," *CPD,* May 6, 1913; "Aviator Sues for Alleged Damages," *ABJ,* July 22, 1913; "Summit County Owns Aeroplanes," *ABJ,* Oct. 6, 1913; "Judgment Confessed," *ABJ,* Oct. 9, 1913; "Aviator Dies as Thousands Look," *CPD,* May 4, 1914; "Sell Equipment; No More Aviation," *ABJ,* July 14, 1914; "Silver Lake Airport"; "New Grounds for Amateur Players," *ABJ,* Mar. 14, 1914.

10. The ideal landing field was described as "about a mile square, the smallest possible permanent field to be a twenty-acre tract clear of trees or other obstructions, one-eighth of a mile wide and one-fourth of a mile-long length in the direction of the prevailing winds. These fields must be well-drained, sandy loam next to level sod being the best ground." *Aircraft Year Book, 1919* (New York: Manufacturers Aircraft Assoc., 1919), 20–21.

11. "University Purchases Stow Flying Field," *DKS,* Oct. 2, 1942; "Expect Zeppelin Industry to Lead," *CPD,* Dec. 11, 1923; "Loses 7 Planes, Keeps His Nerve," *CPD,* Mar. 10, 1929; David Dennis (aviation planner, Ohio Dept. of Transportation), email message to the author, Apr. 17, 2017. An undocumented account has persisted that Reverend Smith's field was used by pilots ferrying World War I "Jenny" planes (i.e., the Curtiss JN-4-D) to other parts of the United States from Canada. The curator at the Glenn H. Curtiss Aviation Museum believes such an undertaking to be highly unlikely given the amount time needed to fly from Canada to Texas with an average Jenny airspeed of 60 mph and a fuel range of only two hours. Also, it would require a large number of experienced pilots, who would then need to return to Canada. Steve Wright, "Airport Flies High After a Slow Start," *DKS,* Apr. 9, 1985; "Kent State May Close Stow Airport," *ABJ,* Aug. 20, 2002; C&S Engineers, Inc., *Kent State University Master Plan, Final Report* (Cleveland, OH: C&S Engineers, 2016), 2:2, https://ksuairportplan.files.wordpress.com/2014/04/ksu-final-report.pdf. "It is our belief that all aircraft were shipped by rail. If the Wings (squadrons) flew the day they arrived in Texas, it would make more sense if the planes were shipped ahead of time, assembled and waiting for the Wing. We also have to remember that there were a large number of 'Jennys' down in Texas for training purposes already on hand in 1917." Rick Leisenring (curator, Glenn H. Curtiss Aviation Museum), email message to the author, Oct. 7, 2016.

12. "Native of Brooke County," *Mill Town Review* (Brooke County, VA), June 28, 1912; "In Southern Prisons: Rev. E. J. Smith Tells His Experience in Them," *ABJ,* Apr. 5, 1893; "Rev. Edward Jackson Smith," in *Akron and Summit County Ohio, 1825–1928,* ed. Scott Dix Kenfield (Chicago: S. J. Clarke, 1928), 3:521; "Celebrates 81st Birthday by Taking Ride in Airplane, Declares It Great," *ABJ,* June 18, 1923; "Summit's Flying Pastor Is Dead," *Massillon (OH) Evening Independent,* Apr. 17, 1930.

13. Roger E. Bilstein, *Flight in America: From the Wrights to the Astronauts,* rev. ed. (Baltimore: Johns Hopkins Univ. Press, 1994), 50–51; Nick A. Komons, *Bonfires to Beacons: Federal Civil Aviation Policy under the Air Commerce Act 1926–1938* (Washington, DC: Federal Aviation Administration, 1978), 18.

14. US Postal Service, *The United States Postal Service: An American History, 1775–2006,* publ. 100 (Washington, DC: US Postal Service), 2007, https://permanent.access.gpo.gov/lps114451/pub100.pdf; Rita Lloyd Moroney, "Airmail," in *Above and Beyond: The Encyclopedia of Aviation and Space Sciences* (Chicago: New Horizons Publishers, 1967), 1:77.

15. *Aircraft Year Book, 1921* (Boston: Small, Maynard), 6–7; Fred Erisman, *Boys' Books, Boys' Dreams, and the Mystique of Flight* (Fort Worth: Texas Christian Univ. Press, 2006), 85; "Free Aeroplane Exhibition Flights at East Market Acres," advertisement, *ABJ,* May 17, 1919; "Silver Lake Highlands Air Stunts that Thrill," advertisement, *ABJ,* Oct. 4, 1919; "Howard Heights See Him Go!," advertisement, *ABJ,* Oct. 10, 1919.

16. Ohio Secretary of State, "New Corporations," in *Department Reports of the State of Ohio.* Columbus, OH: Nemar Publishing, 1920), 120; "An Aviation School Just West of Kent," *R-C,* Apr. 29, 1920.

17. Summit County Fiscal Office, *Grantor-Grantee Index,* s.v. "Edward J. Smith" (book 707, page 266, accessed June 9, 2016); "An Aviation School Just West of Kent"; "Big Aviation Field Being Established West of Kent," *K-Trib,* Apr. 29, 1920; "Ohio Flying Schools, *Aircraft Journal,* May 17, 1920, 9. "Ohio Flying School Buys 5 Planes," *Akron (OH) Times-Press,* Apr. 20, 1920; "Social News," *K-Trib,* Apr. 17, 1924.

18. "Ohio Flying School Buys 5 Planes"; Blower, *Northern Ohio Traction Revisited,* 177; Fred Long, "The Interurban Line," in *Stories from a Stow Native* (Stow, OH: Stow Historical Society, 2001), 181–91; "The Flying Season Is Here," *ABJ,* May 22, 1920.

19. "An Aviation School Just West of Kent"; "Ohio Flying Company Gets Five Planes," *Aerial Age,*

May 10, 1920, 288; "Are War Motors," *ABJ,* July 12, 1920; "Big Flying School to Be Located Near Kent," *Ravenna (OH) Republican,* May 6, 1920; "Classified Advertising," *Aerial Age Weekly,* Sept. 13, 1920, 29; "An Aviation School Just West of Kent."

20. "Air Transportation in Ohio," *Aircraft Journal,* Mar. 13, 1920, 15; "An Aviation School Just West of Kent"; "Aeroplane Flights," advertisement, *ABJ,* July 3, 1920.

21. "An Aviation School Just West of Kent"; "150,000 Attend East Akron Fair," *CPD,* June 18, 1920; William Ganson Rose, *Cleveland: The Making of a City* (1950; repr., Kent, OH: Kent State Univ. Press, 1990), 790; "Progress in America," *Flight,* June 10, 1920, 618; "Air Show to Get Real Start Today," *CPD,* July 3, 1920.

22. "Flying Field Is Given to Akron's Club," *Akron Evening Times,* Apr. 6, 1920; "The National Airway Service Company," advertisement, *ABJ,* Mar. 28, 1920; "Airway Service Co. Formed in Akron," *ABJ,* Feb. 20, 1920; "10,000 at Aerial Field Day Sunday," *ABJ,* June 7, 1920. A later account of this event placed attendance at no more than 500, as people still harbored safety concerns about flight. Harold J. Taylor, "Aviation," *ABJ,* Apr. 9, 1932.

23. James Grant, *The Forgotten Depression: 1921, the Crash That Cured Itself* (New York: Simon and Schuster, 2014); "New Cases," *Akron (OH) Legal News,* Nov. 4, 1921, 1; "Start Aviation Suit," *ABJ,* Dec. 19, 1921.

24. Later accounts about the Ohio Flying School reported that once in bankruptcy, a Mr. Joseph Ash purchased the business, ran the school and field, and then went bankrupt himself in 1923. No person named Joseph Ash nor any credible evidence for this story has been found. Kent State University and TransPlan, Inc., *Master Plan for Kent State University Airport: Andrew W. Paton Field* (New York: TransPlan, 1977), 1:1; "Airport Flies High after a Slow Start."

2. Akron's De Facto Airport, 1923–26

1. *Ohio Fifteenth Federal Census: Statistics of Population by Counties and Minor Civil Divisions, 1930–1920–1910–1900,* compiled by Clarence J. Brown (Columbus, 1930), 7; "Stow," advertisement, *ABJ,* Oct. 4, 1924.

2. "Aviator Here," *K-Trib,* May 3, 1923.

3. "Smith Is Booster for Flying Game," *ABJ,* Apr. 10, 1924.

4. In 1921, these very same bombers, based at Langley Field in Virginia, were used by Mitchell in a series of famous antiship bombing demonstrations. Capt. Walter R. Lawson, who had led those demonstrations, accompanied Mitchell to Stow Field. Mitchell declared him the "greatest bomber officer in the service." "Airmen Gather in Akron Today," *CPD,* Apr. 14, 1923. Shortly after the bombers returned to McCook Field from Stow Field, Lawson's Martin NBS-1 crashed into the Miami River, killing him and two of his crew. The airfield at Fort Benning, Georgia, was named for Lawson, a Georgia native, following his fatal accident. "Four Fliers Die in Fatal Crash at Dayton Field," *Atlanta (GA) Constitution,* Apr. 22, 1923.

5. "Ace Sees Planes Common as Autos," *Huntington (IN) Herald,* June 28, 1923; International News Service, "Ace Sees Planes Common as Autos," *Santa Cruz Evening News,* July 3, 1923; International News Service, "Shops by Airplane," *New Castle (PA) News,* July 14, 1923.

6. "Fred F. Smith Dies; Colorful Politician," *ABJ,* Dec. 4, 1958; "Fred F. Smith," *R-C,* Dec. 4, 1958.

7. Roger E. Bilstein, *Flight in America: From the Wrights to the Astronauts,* rev. ed. (Baltimore: Johns Hopkins Univ. Press, 1994), 52–53; Henry R. Lehrer, *Flying the Beam: Navigating the Early US Airmail Airways, 1917–41* (West Lafayette, IN: Purdue Univ. Press, 2014), 61; "Army Fliers Making Road Maps Visit Akron," *ABJ,* Aug. 22, 1923. The information collected on Stow Field was published by the US Office of the Chief of Air Service in its *Aeronautical Bulletin,* Nov. 1923.

8. Office of the Chief of Air Service, *Aeronautical Bulletin, Number 85, State Series, Ohio: Akron, Stow Field* (Washington, DC: Airways Section, 1923).

9. The "State Summaries Series" expanded coverage by including information on landing facilities for which only summary information was known, as well as for those for which there was a published bulletin. Office of the Chief of Air Service, *Aeronautical Bulletin, Number 34, State Summaries Series, Landing Facilities in Ohio* (Washington, DC: Airways Section, 1925); Office of the Chief of Air Service, *Aeronautical Bulletin, Number 43, Route Information Series: Columbus to Cleveland, Ohio* (Washington, DC: Airways Section, 1923); Office of the Chief of Air Service, *Aeronautical Bulletin, Number 44, Route Information Series: Cleveland, Ohio, to Moundsville, W.VA.* (Washington, DC: Airways Section, 1923).

10. *Aircraft Year Book, 1924* (New York: Aeronautical Chamber of Commerce of America, 1922), 103–11; "FAA Milestones and Events: Origins of the FAA," Federal Aviation Administration, accessed Mar. 22, 2018, https://www.faa.gov/about/history/milestones/media/FAA_Origins.pdf.

11. *Aircraft Year Book, 1923* (New York: Aeronautical Chamber of Commerce, 1923), 125; "Trick Flying

Over Akron Is Prohibited," *ABJ,* Sept. 12, 1923; "Air Traffic Violator Arrested in Sky Race," *New York Times,* Nov. 3, 1923; "Captain Howard [*sic*] Kullberg," Royal Air Force Museum, accessed Nov. 10, 2016, http://www.rafmuseum.org.uk/research/online-exhibitions/americans-in-the-royal-air-force/americans-in-the-british-flying-services-1914–1945/captain-howard-kullberg.aspx.

12. Officers of the Commercial Aircraft Association included Lieutenant Kullberg, president; Fred Smith, treasurer; and V. C. Babcock, director. "Akron Pilots Organize," *CPD,* June 22, 1924; "The Commercial Aircraft Association Plan," *Aviation Week,* June 30, 1924, 699–700; "Commercial Aviators Select Their Officers," *ABJ,* June 19, 1924; "Safety Regulations Adopted by Airmen," *ABJ,* July 25, 1924.

13. "Two Killed as Airplane Falls 1,000 Feet," *ABJ,* June 28, 1924; "Crash Kills Girl and Pilot," *Akron Evening Times,* June 28, 1924.

14. "Sunstroke Caused Crash," *Cincinnati Enquirer,* Aug. 8, 1924.

15. "The Aeronautic Situation Today," *Air Commerce Bulletin* 1, no. 11 (Dec. 2, 1929): 2; "Progress of Commercial Aeronautics in the Past 18 Months," *Air Commerce Bulletin* 2, no. 6 (Sept. 15, 1930): 135; Walter Burton, "Small Airplane Turned out by Akron Inventor, Stands up during Trial Spins," *Akron Evening Times,* July 7, 1923.

16. "Injured Aviator, Unable to Get License, Turns Inventor," *ABJ,* Sept. 25, 1928; Helen Waterhouse, "Wings over Akron," *ABJ,* July 27, 1936. Babcock was one of the original 598 members of the Early Birds of Aviation, which formed in 1928 to record the role of pioneer fliers. Membership in the Early Birds required having soloed as a pilot prior to December 17, 1916, or before up to America's entry into World War I. Babcock qualified based on a flight in Michigan in 1907, when he cracked up taking off in a plane he had designed. He soloed again in 1909 in Seattle in a Wright-type plane he had built himself. "Verne [*sic*] C. Babcock, 1886–1972," Early Birds of Aviation, Inc., accessed Oct. 31, 2016, http://earlyaviators.com/ebabcock.htm.

17. Babcock had been a designer with a short-lived company called Advance Aircraft Corp. in Los Angeles, California, in 1916. That firm reportedly outfitted a biplane with a Stanley Steamer engine. This effort to use steam power proved unsuccessful, and the company was dissolved. "Advance Aircraft Corp.," *San Francisco Chronicle,* Oct. 14, 1916; "Stock Permits Revoked," *San Francisco Chronicle,* Mar. 29, 1917.

18. "Babcock, Babcock-Vlchek," Aerofiles, last modified Mar. 11, 2009, http://www.aerofiles.com/_ba.html; Paul H. Rear, "Airports and Airways: Akron News," *Aviation Week,* Dec. 8, 1924, 1372; "New Altitude Record Set," *Akron (OH) Times-Press,* Apr. 13, 1925.

19. "Reliability Tours Land Public Trust," *The Henry Ford,* last modified Aug. 3, 2012, https://www.thehenryford.org/explore/blog/reliability-tours-land-public-trust; "25 Planes Are Ready for 2,500 Mile Tour," *Detroit Free Press,* Aug. 7, 1926; Lesley Forden, "Second Tour—Bigger, Better," in *Ford Air Tours, 1925–1931* (New Brighton, MN: Aviation Foundation of America, 1972), 25; "Walter Beach Named Winner of Flying Tour," *Detroit Free Press,* Aug. 22, 1926.

20. "Commissioner Fred Smith Asked to Resign," *Akron (OH) Times-Press,* Oct. 24, 1925.

21. "Famous Ace Commends Beacon Journal Model Plane Contest," *ABJ,* May 29, 1925; "Army Birdmen to Appear over City," *ABJ,* May 23, 1924; "Pilots and Some Planes Which Will Take Part in Aerial Circus over Akron," *ABJ,* May 21, 1924; "School Children Told How Planes Are Built, Flown," *ABJ,* May 24, 1924.

22. "Air's Dare Devils 'Do Their Stuff' as 75,000 Watch," *CPD,* June 8, 1925; "200 Akron School Boys to Compete in Beacon Journal's Second Annual Model Aircraft Meet," *ABJ,* May 27, 1925; "Offer 20 Prizes to Boys in Beacon Journal Plane Contest," *ABJ,* June 1, 1925.

23. "The First 'Goodyear Blimp'—Pilgrim of 1925," Airships.net: A Dirigible and Zeppelin History Site, accessed Sept. 26, 2015, http://www.airships.net/goodyear-blimp/goodyear-blimp-pilgrim.

24. "Morton & Maxson Declared Winners of Balloon Race," *Wingfoot Clan,* July 29, 1925.

25. "Plan Another Air Meeting at Stow," *K-Trib,* June 25, 1925; "Fast Aeroplanes Will Participate in Flying Meet," *ABJ,* July 13, 1925; "One Balloon Out of Four in Akron Race Still Going," *CPD,* July 20, 1925.

26. "Balloon Contest Today Heralds Akron Centenary," *CPD,* July 18, 1925; "Plan to Deliver Complete Motor Car by Airplane," *ABJ,* July 16, 1925; "One Balloon Out of Four in Akron Race Still Going"; "Tie Race May be Result of Balloon Flight," *ABJ,* July 20, 1925; Traffic Deputy, letter to the editor, *ABJ,* Aug. 4, 1925; D. B. Goss, letter to the editor, *ABJ,* July 22, 1925; Boyd M. Hause, letter to the editor, *ABJ,* Aug. 12, 1925.

27. *Aircraft Year Book, 1926* (New York: Aeronautical Chamber of Commerce, 1926), 63.

28. "Doubts Recovery of $468,091 in Road Scandal Case," *ABJ,* Jan. 16, 1925; "Fred Smith, 6 Other Men, Indicted," *ABJ,* Apr. 4, 1925; "Dan Stein, in Poor Health, Leaps to Death off North Hill Viaduct," *ABJ,* July 24, 1925; "Attack Legality of County Road Case

Indictments," *ABJ*, Apr. 24, 1925; "Booth Balks Grand Jury's Compromise on Roads," *ABJ*, July 3, 1925.

29. "Classifieds—Real Estate," *ABJ*, Apr. 29, 1925.

30. "Road Case Grand Jury is Illegal," *ABJ*, July 17, 1925; "Smith Is Ready to Resign Office," *ABJ*, Oct. 24, 1925; "Speculation on Successor for Smith Started," *ABJ*, Oct. 27, 1925.

31. "The Fageol Safety Coach—A Breakthrough in Bus Design," *Old Motor*, last modified Jan. 15, 2014, http://theoldmotor.com/?p=111779; "Stow Field May Become Pivot of Airplane Traffic," *ABJ*, June 22, 1926.

32. "Factory to Build Airplanes Will Be Erected at Stow," *ABJ*, June 25, 1926; "Plane Manufacturers Get Stow Field," *K-Trib*, May 20, 1926.

33. Jung's previous experience included building airplanes, instructing, and flying a three-engine airplane for the London-Berlin airline route. He was hired by Safety Airways as a research pilot. "German Ace Is Engineer for Builders Here," *K-Trib*, July 1, 1926; "Plane Manufacturers Get Stow Field."

34. "They're Off," *Cincinnati Enquirer*, May 31, 1927; "Thrilling Descent on Two Parachutes," *ABJ*, Aug. 29, 1927; "Parachute Record Set," *CPD*, June 28, 1926; "The Human Bullet Parachute Jump," advertisement, *ABJ*, July 10, 1926; "Will Again Attempt New 'Chute' Record," *ABJ*, July 17, 1926.

35. "Night Flying Circus," advertisement, *ABJ*, July 30, 1926; "Exhibit of Night Flying at Stowe," *Ravenna (OH) Republican*, July 27, 1926; "Flaming Plane Falls at Stow; Pilot Is Dying," *ABJ*, Aug. 9, 1926.

36. "Cause of Air Crash Unknown," *K-Trib*, Aug. 12, 1926; "Jung Back in Kent, Gaining," *K-Trib*, Sept. 30, 1926.

37. "Plane Crashes at Stow," *K-Trib*, Sept. 2, 1926; "Fly with Esch," advertisement, *ABJ*, Oct. 1, 1926; "Want Ads," *K-Trib*, Nov. 11, 1926.

38. *Aircraft Year Book, 1920* (New York: Doubleday, 1920), 66; Janet R. Daly Bednarek, *America's Airports: Airfield Development, 1918–1947* (College Station: Texas A&M Univ. Press, 2001), 5.

39. Bednarek, *America's Airports*, 6; Nick A. Komons, *Bonfires to Beacons: Federal Civil Aviation Policy under the Air Commerce Act 1926–1938* (Washington, DC: Federal Aviation Administration, 1978), 173–75.

40. "Akron May Be Made Center in Ohio of Air Mail Transportation," *Portsmouth (OH) Daily Times*, July 9, 1924.

41. "Akron to Bid for Air Mail Service," *CPD*, Sept. 23, 1924; "Two Mail Pilots Risk Lives Here," *CPD*, Sept. 21, 1924.

42. "Cleveland-Hopkins International Airport," *Encyclopedia of Cleveland History*, last modified July 14, 1997, http://ech.case.edu/cgi/article.pl?id=CIA; "Cleveland-Hopkins Airport—History," Cleveland Hopkins International Airport, accessed Oct. 23, 2016, http://www.clevelandairport.com/company/history; "Airport List," *CPD*, Sept. 9, 1925.

43. "Expect Zeppelin Industry to Lead," *CPD*, Dec. 11, 1923; "To Urge Stow as Zeppelin Center," *K-Trib*, Dec. 20, 1923.

44. "Akron Asks 50–50 Mail and Zep Site," *CPD*, Oct. 29, 1924.

3. Years of Promise, 1927–29

1. "FAA Milestones," *Federal Aviation Administration*, last modified Jan. 3, 2018, https://www.faa.gov/about/history/milestones/media/FAA_Origins.pdf; "Owner of Stow Field Will Sell," *CPD*, Jan. 15, 1927.

2. Frederic J. Haskin, "Haskin's Letter," *Evening Review* (East Liverpool, OH), Dec. 16, 1926; "Aviation Field at Stow Leased to New Concern," *ABJ*, Apr. 6, 1927.

3. "Quick Air Trip," *Hudson (OH) Herald*, June 24, 1926; Schetter, "Rubber City Plane Pilots," *ABJ*, Sept. 19, 1928; "Loses 7 Planes, Keeps His Nerve."

4. "Loses 7 Planes, Keeps His Nerve," *CPD*, Mar. 10, 1929; "Board Suspends License of Hugh C. Robbins," *Civil Aeronautics Journal* 2, no. 20 (1941): 259; "Inventions Wanted," *Popular Science* 146, no. 5 (May 1945): 55; Norman Frisch, "Gold Mine in the Sky," *Salute* 12, no. 11 (Nov. 1947): 5; Clyde E. Schetter, "Rubber City Plane Pilots," *ABJ*, Sept. 19, 1928; Hugh C. Robbins, US Patent 3,029,581, issued Apr. 17, 1962; Hugh C. Robbins, US Patent 3,047993, issued Aug. 7, 1962.

5. "Aviation Field at Stow Leased to New Concern."

6. "Aviator at New Stow Field," *Hudson (OH) Herald*, Apr. 7, 1927.

7. US Department of Commerce, *Airway Bulletin, Number 86: Ohio-Akron, Stow Field* (Washington, DC: Aeronautics Branch, 1927), 1–2.

8. "Double Parachute Jump Stow Feature," *ABJ*, July 20, 1927; "Boden, Former Daredevil, Is Near Death in Hospital," *ABJ*, Jan. 15, 1937.

9. "He Has Seen Lindy Operate as Pilot," *K-Trib*, June 9, 1927; "New Plane and a New Pilot in Stow," *K-Trib*, June 30, 1927.

10. The Embry-Riddle Company, formed in Cincinnati in 1925, merged with the Aviation Corporation (AVCO) in 1929 and phased out its flying school. AVCO would become American Airways and then American Airlines. The Embry-Riddle flight school was begun anew in 1939 in Florida by one of the original founders. "The Early Years," Embry-Riddle

Aeronautical University, accessed Oct. 15, 2016, http://erau.edu/administration/archives/early-years; "Aerial Passenger Service for City Starts Thursday," *ABJ*, May 15, 1927.

11. "Aviators Battle Bad Weather as Air Line Opens," *ABJ*, May 19, 1927; "Ask 100 to Ship by New Air Express to Ohio and South," *CPD*, May 18, 1927; "Air Service Fails to Prove Profitable," *ABJ*, Sept. 24, 1927.

12. Lindbergh was a former barnstormer and airmail pilot attracted to the $25,000 prize offered for the first flier or group of fliers to make a nonstop transatlantic flight in a heavier-than-air craft. Lindbergh achieved the feat on May 21, 1927, in a thirty-three-and-a-half-hour flight from New York's Roosevelt Field to Le Bourget airfield in Paris. James E. Crehan and Tim Brady, "Developments between the Wars," *The American Aviation Experience: A History*, ed. Tim Brady (Carbondale: Southern Illinois Univ. Press, 2000), 154–55; *Aircraft Year Book, 1928* (New York: Aeronautical Chamber of Commerce, 1928), 3.

13. "Reporter, Stow Flier to Greet Lindbergh," *ABJ*, Aug. 1, 1927; "'Lindy' Circles City, Dropping His Greetings," *ABJ*, Aug. 3, 1927.

14. "Airplanes Will Open Falls Cleanup Drive," *ABJ*, Apr. 25, 1929; Harold J. Taylor, "Aviation," *ABJ*, Aug. 28, 1930. Robbins's airplane flew two nights over Akron and surrounding areas to mark the opening of the *Akron Beacon Journal*'s outdoor Christmas lighting contest. The red and green lights were supplied by the Northern Ohio Power and Light Company, the batteries by Firestone Tire and Rubber Company, and the fuel by Standard Oil Company. "Christmas Airplane to Fly over Greater Akron Two Nights," *ABJ*, Dec. 11, 1929.

15. Schetter, "Rubber City Plane Pilots," *ABJ*, Sept. 19, 1928; Advance Aircraft Company changed its name to Waco Aircraft Company in June 1929 to more closely align the company name with their product. Raymond H. Brandly, *WACO Airplanes: "Ask Any Pilot"* (Dayton, OH: R. H. Brandly, 1979), 67; "Loses 7 Planes, Keeps His Nerve"; Michael Rotman and F. X. O'Grady, "Cleveland National Air Races," *Cleveland Historical*, accessed Mar. 25, 2018, https://clevelandhistorical.org/items/show/151; "Robbins Takes More Space at Cleveland," *ABJ*, May 18, 1929.

16. "Survey Reveals Airport Fever Is Sweeping Country," *CPD*, July 9, 1927.

17. Ryan DuBosar, "Darrowville Only a Memory to Most Residents," *Stow (OH) Sentry*, Aug. 11, 1991; "Air Board Meets to Organize for Survey of Fields," *ABJ*, July 22, 1927; "Seven Airport Sites Offered City of Akron," *ABJ*, Aug. 3, 1927.

18. "Four Fields Are Recommended by Site Committee," *ABJ*, Aug. 23, 1927; "Pick Darrowville Land for Airport," *CPD*, Aug. 23, 1927; "Voters to Act on $1,135,000 Bond Measures," *ABJ*, Sept. 7, 1927.

19. "Air Mail Hinges on Fate of Bond Issue for Port," *ABJ*, Oct. 11, 1927. On June 4, 1927, Clarence Chamberlin flew from Long Island, New York, to Eiseleben, Germany, setting the record both for piloting the first transatlantic passenger flight and for establishing a distance flight record of over 3,900 miles. *Aircraft Year Book, 1928*, 21; "Chamberlin Due at Akron Today," *CPD*, Nov. 7, 1927.

20. "Vote on Special Issues," *ABJ*, Nov. 9, 1927; "Navy Officials May Decide Soon about Dirigibles," *ABJ*, Dec. 15, 1927.

21. "Form Airport Firm at Stow," *K-Trib*, Nov. 17, 1927; "Plan Extensions of Flying Service," *ABJ*, Nov. 10, 1927; "More Hangars at Stow," *K-Trib*, Dec. 15, 1927.

22. "Want Zeppelin Hangars at Stow," *K-Trib*, Dec. 29, 1927; "Civic Bodies May Raise Funds for Airport at Stow," *ABJ*, Dec. 30, 1927.

23. "Proposed Air Field Plan to Cost $100,000," *ABJ*, Jan. 12, 1928; "Urges Issue of Bonds for Stow Airport," *R-C*, Jan. 17, 1928; Donald B. Holmes, *Air Mail: An Illustrated History, 1793–1981* (New York: Clarkson N. Potter, 1981), 124–25.

24. "Seek Financial Aid to Improve Airport at Stow," *ABJ*, May 10, 1928; "Officials of Stow Field Lay Program for Summer Work," *R-C*, Mar. 15, 1928.

25. "Akron (OH) Ready to Battle for Big Airport," *ABJ*, Feb. 18, 1928; "Diversion of Funds Planned for Airport," *ABJ*, Feb. 11, 1928.

26. "Extend Options upon Proposed Site of Airport," *ABJ*, June 19, 1928; "Plan Test Flights on New Mail Line," *ABJ*, June 18, 1928.

27. "Set to Start Zep Work at Akron," *CPD*, Oct. 28, 1928.

28. "Three Apply for Airport Hangar Space," *ABJ*, Nov. 29, 1928; "Airport Needs for Year Total $12,700," *ABJ*, Dec. 21, 1928; "Work on Airport Hangars Will Be Started at Once," *ABJ*, May 2, 1929.

29. "Name Sales Head of Robbins School," *ABJ*, Jan. 21, 1929; "Plan Your Career in Aviation," advertisement, *ABJ*, Mar. 25, 1929; "Boys Are Ready for Model Plane Race at Airport," *ABJ*, May 31, 1929.

30. Schetter, "Rubber City Plane Pilots," *ABJ*, Sept. 19, 1928; Clyde E. Schetter, "Rubber City Plane Pilots," *ABJ*, Sept. 24, 1928; Clyde E. Schetter, "Rubber City Plane Pilots," *ABJ*, Sept. 25, 1928; "Classified Advertisements: Wanted," *Aero-Digest*, June 1929, 289, accessed Dec. 11, 2015, https://archive.org/stream/aerodigest1415unse#page/n567/mode/2up.

31. Douglas Belfield, "A Chat with Flyer Frankie

Renner," *Stow (OH) Citizen*, Sept. 1, 1977; "Frankie Renner, 1897–1985," *Akron Women's History*, accessed Aug. 23, 2015, https://blogs.uakron.edu/womens history/2013/09/09/frankie-renner-1897–1985/; "Akron Girl Flier Fails to Set Mark," *ABJ*, Mar. 26, 1931; "Akron Girls Form Flying Sorority," *Daily Times* (New Philadelphia, OH), Mar. 25, 1932; "Frankie Renner, 1897–1985"; Harold J. Taylor, "Aviation," *ABJ*, Apr. 14, 1932; "Frankie Renner," *Broadcasters Hall of Fame*, accessed Oct. 20, 2015, http://www.broadcastershallof fame.com/biofrenner.htm; "Miss Frankie G. Renner," *ABJ*, May 28, 1985.

32. "She Knows Her Plane as She Does Her Typewriter," *CPD*, July 22, 1928; "Buckeye State Air Flashes," *CPD*, Oct. 27, 1929; "111 Women in U.S. Licensed to Fly Planes and 4 Are from Ohio," *CPD*, Nov. 6, 1929; Harold J. Taylor, "Aviation," *ABJ*, Nov. 11, 1929, Dec. 17, 1929.

33. "Freddie Lund Will Perform Daring Feats at Stow Field," *ABJ*, Nov. 22, 1928; "Freddie Lund Stunts over Stow Airport," *ABJ*, Nov. 23, 1928. Lund did perform the dangerous outside loop at the National Air Races in Cleveland in August 1929.

34. "$50,000 Damage Estimated after Plane Port Fire," *ABJ*, Mar. 7, 1929; "Eight Planes, Hangar Burn at $50,000 Loss," *Akron (OH) Times-Press*, Mar. 7, 1929.

35. "Stow Field Hangar Being Constructed," *Cuyahoga Falls (OH) Reporter*, Mar. 29, 1929; Robbins's replacement airplanes following the fire included a Waco 90, a 165-horsepower J-5 Waco, a 220-horsepower J-5 Waco, and a 150-horsepower Waco. Harold J. Taylor, "Aviation," *ABJ*, July 30, 1929.

36. "Civic Leaders Ready to Work for New Industries," *ABJ*, July 25, 1928; "Foundation to Back College of Aviation," *ABJ*, Jan. 18, 1929.

37. Support was given to build a hangar at the new municipal field, costing $40,000 to $50,000. An additional $25,000 was provided by the foundation. "Board Passes on Plans for Akron Airport Hangar," *ABJ*, Feb. 13, 1929; "Foundation Acts to Approve Financing," *ABJ*, June 13, 1929.

38. Air Services expanded operations beyond instruction to become more of a fixed-base operator selling airplanes and offering taxi service and rides. "Air Services, Incorporated," advertisement, *ABJ*, May 27, 1929.

39. "Lindy's Approval of Akron Flying School Is Sought," *ABJ*, Jan. 21, 1929; "Akron Welcomes Amelia," *ABJ*, Feb. 15, 1929; "City Thrilled by Air Circus of Army, Navy," *ABJ*, Aug. 8, 1931. Earhart's final visit to the area came after her 1932 solo flight across the Atlantic. She spoke at Kent State on "Flying for Fun" on October 10, 1934, as part of a nationwide speaking tour. "Noted Aviatrix to Appear Here October 10," *DKS*, Oct. 4, 1934.

40. Helen Waterhouse, "Akron Will Make Bid for National Air Races," *ABJ*, Nov. 28, 1940.

41. "Mid-City Airport Owners Take out Delaware Papers," *ABJ*, Nov. 14, 1929; "Maj. Lanphier Heads Airport," *Detroit Free Press*, Nov. 17, 1929.

42. Harold J. Taylor, "Mid-City Airport Being Completed by Carl Wollam," *ABJ*, July 31, 1929; "Big Airport at Darrowville Promises Stupendous Growth," *Hudson (OH) Herald*, Oct. 10, 1929; Harold J. Taylor, "Aviation," *ABJ*, Sept. 20, 1929.

43. Harold J. Taylor, "Aviation," *ABJ*, Oct. 25, 1929.

44. "Visitors at Local Airport Fail to Recognize Colonel," *ABJ*, Dec. 28, 1929; Waterhouse, "Akron Will Make Bid for National Air Races." The apocryphal account of a visit by Lindbergh to Stow Field first appeared in writing in 1976 and was regularly restated. The story often includes a visit by American aviator Floyd Bennett, also unsubstantiated. The following are just a few of the accounts. Eugene G. Ripple, "History of Aerospace Technology at Kent State University, 1941–1976" (draft master's thesis, Municipal University of Akron, Mar. 1976?), 40, KSUSCA; Kent State University and TransPlan, Inc., *Master Plan for Kent State University Airport: Andrew W. Paton Field* (New York: TransPlan, 1977), 1:1; Steve Wright, "Airport Flies High after a Slow Start, *DKS*, Apr. 9, 1985; C&S Engineers, Inc., *Kent State University Master Plan, Final Report* (Cleveland, OH: C&S Engineers, 2016), 2:2, https://ksuairportplan.files.wordpress.com/2014/04/ksu-final-report.pdf.

45. W. C. Howells, "'Gypsy' Aviators Outlawed in Ohio," *CPD*, June 26, 1929; "FAA Historical Chronology," accessed July 24, 2018, https://www.faa.gov/about/history/chronolog_history/media/b-chron.pdf.

46. Howells, "'Gypsy' Aviators"; "Ohio Assembly Will Be Asked to Adopt Federal Air Code for State," *CPD*, Apr. 22, 1928; Robert F. Kirk, *Flying the Lindbergh Line: Then and Now* (Bloomington, IN: Authorhouse, 2013), 46.

47. Jake Falstaff, "A Voice from the Gallery," *ABJ*, May 5, 1928; "Mayor Wants Paint for 'Salem' Air Sign," *Salem (OH) News*, May 7, 1928; "Ohio Villages Call Rogers' Paint Bet," *CPD*, May 4, 1928.

48. Fred Smith sent his thanks on behalf of Stow Field to the Kent Chamber of Commerce for their role in getting the job done. James D. Hartshorne, "New Air-Rail Hook-Up Marks Another Milestone in Aviation," *CPD*, June 3, 1928; "Buckeye State Air Flashes," *CPD*, Nov. 17, 1929. Redmond's Studebaker garage was at 330 North River Street. This street is

now called Gougler, and the property is vacant land. "'Kent' on Roof Lights the Way for Aviators," *K-Trib*, Feb. 23, 1928; "Smith Thanks Kent for Air-Marking," *R-C*, Nov. 16, 1929.

49. Eugene F. Gleason, "Doolittle Raider Pilots Cleveland Merchants' War Bond Drive Today," *CPD*, July 1, 1942; "Col. William Marsh 'Bill' Bower," *Daily Camera* (Boulder, CO), Jan. 11, 2011, http://www.dailycamera.com/obits/ci_17057961; T. Rees Shapiro, "Bill Bower, Last Surviving Bomber Pilot of WWII Doolittle Raid, Dies at 93," *Washington Post*, Jan. 15, 2011.

50. Milt Freudenheim, "President's Pilot Had Early Training," *Corpus Christie (TX) Caller-Times*, Nov. 15, 1953; Mark J. Price, "Falls Aviator Coined 'Air Force One,'" *ABJ*, Nov. 24, 2014.

51. John McMillion, "Akron Twins among Early Kings of the Skies," *ABJ*, Sept. 20, 1987; John Durbin, "Twin Brothers Develop New Spin-Proof Airplane," *Times Recorder* (Zanesville, OH), July 5, 1940; "Funks Determine to Stay in Akron," *ABJ*, Apr. 7, 1939; "Hearing Friday on Funk Factory," *ABJ*, June 20, 1941; Coffeyville Aviation Heritage Museum, accessed July 19, 2017, http://www.kansastravel.org/coffeyvilleaviationheritagemuseum.htm.

4. The Depression Sets In, 1930–38

1. Robert P. Lamont, "Preening Wings of Commerce," *Air Commerce Bulletin* 1, no. 3 (Aug. 1, 1929): 1; Clarence M. Young, "Civil Aviation and the Trend of Development," *Air Commerce Bulletin* 2, no. 9 (Nov. 1, 1930): 215.

2. "Air Commerce," *Twenty-Second Annual Report of the Secretary of Commerce* (Washington, DC: US Government Printing Office, 1934), xxii; "Status of Federal Airways System on June 30 of Years Indicated," *Air Commerce Bulletin* 9, no. 3 (Sept. 15, 1937): 50; "Progress of Civil Aeronautics in the United States, 1927–1936," *Air Commerce Bulletin* 8, no. 12 (June 15, 1937): 263; Nick A. Komans, *Bonfires to Beacons* (Washington, DC: Federal Aviation Administration, 1978), 223.

3. Komans, *Bonfires to Beacons*, 222; "Progress of Civil Aeronautics in the United States," 265.

4. "Goodyear Plant Institutes Four Six-Hour Shifts," *ABJ*, Oct. 14, 1930; "Sale of Apples by Unemployed Starts in Akron," *ABJ*, Dec. 2, 1930; "Seek $1,000 Fund for Buying Seeds," *ABJ*, May 12, 1931.

5. "Aviation Instruction," classified advertisement, *ABJ*, Jan. 22, 1930; "Aviation Instruction," classified advertisement, *ABJ*, Feb. 4, 1930; "Aviation Instruction," classified advertisement, *ABJ*, Feb. 17, 1930.

6. Harold J. Taylor, "Aviation," *ABJ*, Jan. 16, 1930; Donald E. Kehoe, "Aviation Training," *Popular Aviation* (Nov. 1928), 60; "Plan Clubs to Cut High Cost of Flying," *New York Times*.

7. Harold J. Taylor, "Aviation," *ABJ*, Nov. 4, 1930, Dec. 20, 1930; "Free Airplane Ride with New Aviation Preparatory Course," advertisement, *ABJ*, Apr. 7, 1931.

8. Lauren D. Lyman, "Gliding Flights Spur Nation-wide Interest," *New York Times*, Apr. 13, 1930.

9. Merrill Hamburg, "First of Airplane Series Tells How to Build Glider," *ABJ*, Feb. 25, 1929; "Akron off to Gliding Start," *CPD*, Apr. 7, 1929; "Aviation Instruction," classified advertisement, *ABJ*, Apr. 26, 1930; Harold J. Taylor, "Aviation," *ABJ*, May 21, 1930.

10. "Three Airports Boast of Akron," *Shamokin (PA) News Dispatch*, Apr. 3, 1931; "Harold J. Taylor, "Aviation," *ABJ*, Apr. 1, 1930; US Department of Commerce, *Airway Bulletin Number 86, rev.: Ohio-Akron (Cuyahoga Falls), Stow Field* (Washington, DC: Aeronautics Branch, 1931), 1.

11. "Flying Minister Dies in Stow at Age of 87 Years," *ABJ*, Apr. 17, 1930.

12. "H. C. Robbins to Take Part in New Movie," *ABJ*, Aug. 27, 1931; "Dick Grace," *IMDB*, accessed Jan. 17, 2017, http://www.imdb.com; "Dick Grace," advertisement, *CPD*, Apr. 5, 1930; "Aviation Students Hear Movie Flier," *ABJ*, Sept. 16, 1930.

13. Harold J. Taylor, "Aviation," *ABJ*, June 24, 1930; "Keeping up on Air Events," *CPD*, July 20, 1930; "Women Plan Stunts for Aerial Festival," *CPD*, Aug. 31, 1930.

14. Harold J. Taylor, "Aviation," *ABJ*, Jan. 31, 1931; Andrew Heins, National WACO Club president, email message to the author, Mar. 3, 2017; "Akron Woman to Seek New Mark," *Evening Independent* (Massillon, OH), Jan. 31, 1931; H. Parker Lowell, "Akron Aviatrix to Try for Altitude Record," *Akron (OH) Times-Press*, Feb. 15, 1931.

15. Harold J. Taylor, "Aviation," *ABJ*, Feb. 10, 1931; "Woman to Seek Altitude Mark," *Salem (OH) News*, Feb. 10, 1931; "Aviator Schroeder Falls Five Miles, Frozen, Unconscious, Lives," *Sandusky (OH) Register*, Feb. 28, 1920.

16. Harold J. Taylor, "Aviation," *ABJ*, Feb. 27, 1931, Mar. 5, 1931; "Miss Renner Seeks New Altitude Mark," *ABJ*, Mar. 13, 1931; "Cleveland Girl Flyer Seeks Mark," *Coshocton (OH) Tribune*, Mar. 13, 1931; "A Chat with Flyer Frankie Renner," *Stow (OH) Citizen*, Sept. 1, 1977.

17. Frankie Renner Scrapbook, Renner Collection, International Women's Air and Space Museum, Cleveland, Ohio.

18. "Cleveland Girl Flyer Seeks Mark"; "Miss Renner Seeks New Altitude Mark."

19. "Akron Woman Aviator Goes up Six Miles," *ABJ*, Mar. 14, 1931.

20. Harold J. Taylor, "Aviation," *ABJ*, Apr. 22, 1931; "Akron Girl Flier Fails to Set Mark," *ABJ*, Mar. 25, 1931; "Frankie Renner to Try Again for Altitude Record," *Coshocton (OH) Tribune*, Apr. 12, 1931; Harold J. Taylor, "Aviation," *ABJ*, May 26, 1931; "Find Music in Plane Motor's Hum," *CPD*, Jan. 2, 1932.

21. "Stow Field Pilots Give Aerial Stunts," *ABJ*, July 27, 1931; "Business Opportunities," classified advertisement, *ABJ*, Oct. 8, 1931.

22. Harold J. Taylor, "Aviation," *ABJ*, Jan. 2, 1932, Apr. 20, 1932, June 15, 1932.

23. "Robbins Abandons Stow Flying Field," *ABJ*, June 30, 1932; "A Chat with Flyer Frankie Renner"; "Two Will Appraise Airport Property," *ABJ*, July 1, 1932.

24. "Stow Flying Field Is Granted Charter," *ABJ*, July 18, 1932. Lauretta Schimmoler was the first female airport manager in Ohio, beginning at Bucyrus airport in July 1931. "Castalia Pilot Is Forced down in Bucyrus Race," *News-Messenger* (Fremont, OH), July 20, 1931; Harold J. Taylor, "Aviation," *ABJ*, July 19, 1932, July 28, 1932; "Business Opportunity," classified advertisement, *ABJ*, Aug. 12, 1932.

25. "Party Saluting Al's 90 Years Like Script out of Hollywood," *ABJ*, May 3, 1977; Rebecca Maksel, "The 'Squadron of Death' That Performed in High Heels," *Air & Space Magazine*, Feb. 25, 2014, http://www.airspacemag.com/daily-planet/squadron-death-180949714; Margaret Schippel Forest, letter to the editor, *ABJ*, June 13, 1977.

26. "Akron District Aviatrices Plan 'Squadron of Death,'" *ABJ*, Mar. 21, 1932; "Squadron of Death Selects Black Crow as Its Mascot," *ABJ*, May 14, 1932; Harold J. Taylor, "Aviation," *ABJ*, Mar. 24, 1932, June 23, 1932.

27. Harold J. Taylor, "Aviation," *ABJ*, Jan. 27, 1933, Apr. 20, 1933; "Airport and Airway Changes and Warnings," *Air Commerce Bulletin* 4, no. 22 (May 15, 1933): 576.

28. "Flames Sweep Stow Airport," *ABJ*, Aug. 14, 1933; Dorothy Doran, "Radio Draws Names Back to Air Waves as Fall Approaches," *ABJ*, Aug. 18, 1933.

29. Joseph J. Corn, *The Winged Gospel: America's Romance with Aviation, 1900–1950* (New York: Oxford Univ. Press, 1983), 71–73; Dydia DeLyser, "Flying: Feminisms and Mobilities—Crusading for Aviation in the 1920s," in *Geographies of Mobilities: Practices, Spaces, Subjects*, ed. Tim Cresswell and Peter Merriman (Burlington, VT: Ashgate, 2011), 83–88.

30. "General Airway Information," *Airway Bulletin* 23, no. 1 (Sept. 1, 1932): 66.

31. Summit County Court of Common Pleas, Sheriff's Deed, no. 394760, July 16, 1936; Joseph C. Van Devere (son, Rudy Van Devere), phone conversation with the author, Nov. 29, 2016.

32. Gail Mishler was chief pilot. Management included Joe Conti, *Beacon Journal* zinc engraver, B. M. Jennings of North Carolina, and James H. Woodall of Cuyahoga Falls, who was also the airport operator. An "all-metal" training airplane was added in August, and one equipped for "blind flying," or for flying by instruments only, was added the next spring. Rides on a Ford Tri-Motor airplane were available. By fall of 1937, Mishler left with plans to fly for China in the Chinese-Japanese war and was replaced by Art Laurence. Helen Waterhouse, "Wings over Akron," *ABJ*, Mar. 29, 1937, Apr. 12, 1937. The first "blind flight" relying on instruments only for navigation was made by Lt. James Doolittle on Sept. 24, 1929. Lauren D. Lyman, "Landing by Instrument Means New Safety in Air," *New York Times*, Sept. 29, 1929; Waterhouse, "Wings over Akron," Aug. 16, 1937, Apr. 4, 1938; "Personals," *Journal-Gazette* (Mattoon, IL), Sept. 20, 1937; Waterhouse, "Wings over Akron," Oct. 18, 1937, Nov. 8, 1937.

33. Herzel H. E. Plaine, "State Aviation Legislation," *Journal of Air Law and Commerce* 14, no. 3 (1947): 335, http://scholar.smu.edu/jalc/vol14/iss3/6.

34. Waterhouse, "Wings over Akron," Nov. 14, 1938. Although the titles of airport manager and airport director were both used, airport manager will be used throughout this book.

35. Harold J. Taylor, "Aviation," *ABJ*, June 30, 1931; "Ohio Optometrists to See Air Display," *ABJ*, Aug. 9, 1930; "Fly to the Games," advertisement, *ABJ*, Oct. 2, 1930; "Stunt Glamour Thrills Crowd Dedicating Port," *ABJ*, June 16, 1931; "Terminal Building Modern Gateway to Port of Akron," *ABJ*, June 15, 1931; "Akron Welcomes Mrs. Hoover," *ABJ*, Aug. 8, 1931; "Greater Days to Come," editorial, *ABJ*, Aug. 10, 1931.

36. C. W. Howard, "Sparks Appeals for U.S. Aid," *ABJ*, May 4, 1933; "Causes Leading to Reduced Services Outlined by Mayor," *ABJ*, May 6, 1933; "Council Split over Slashes," *ABJ*, May 6, 1933; "$7,450,000 Works Program Wins, Light Plant Approved," *ABJ*, Nov. 8, 1933; "CWA Work Improves Airports in Ohio," *ABJ*, Dec. 12, 1933.

37. "Air Services, Inc., Refinancing Seen," *ABJ*, Apr. 16, 1935; "Reject Proposals of Air Services," *ABJ*, Nov. 21, 1935; "Wells is Receiver for Air Services," *ABJ*, Apr. 7, 1936; "Betz, Van Devor [*sic*] Get Air Services," *ABJ*, Apr. 27, 1936; Lloyd Stoyer, "Just Plane Lucky He's Still Alive," *ABJ*, Mar. 9, 1958; "Articles of Incorporation of Akron Airways, Inc.," *Ohio Secretary of*

State, July 14, 1936, http://www2.sos.state.oh.us/reports/rwservlet?imgcl2g&Din=C048_0043.

38. Stoyer, "Just Plane Lucky He's Still Alive."

39. Stoyer, "Just Plane Lucky He's Still Alive"; "Van Devere is Auto-Plane Man," *ABJ,* Feb. 6, 1951; "Caterpillar Club," *51 Squadron,* accessed Feb. 7, 2017, http://www.51squadron.com/CaterpillarClub.html; "CAK Authority Dedicates Aviation Park and Celebrates Centennial Anniversary of Flight," Akron-Canton Airport, Dec. 17, 2003, accessed Mar. 17, 2016, http://www.akroncantonairport.com/newsroom/2006/02/cak-authority-dedicates-aviation-park-and-celebrates-centennial-anniversary-of-flight.

40. Helen Waterhouse, "Wings over Akron," *ABJ,* Aug. 23, 1937, Nov. 17, 1937; "FLY," advertisement, *ABJ,* July 24, 1937; Helen Waterhouse, "Wings over Akron," *ABJ,* June 6, 1938.

41. "Akron Airways Loses Hangar at Airport," *ABJ,* Aug. 31, 1938; "Lone Airport Bid Accepted by City," *ABJ,* Dec. 1, 1938.

42. "Issoudun Mid-City Flying Corporation Announces," advertisement, *ABJ,* Sept. 7, 1929; "Mid-City Aviation Company Expands," *ABJ,* Mar. 6, 1930; "Hudson Hears It's to Be Air Center," *CPD,* July 30, 1930.

43. Harold J. Taylor, "Aviation," *ABJ,* Nov. 23, 1931; "School of Flying Leaves Mid City," *ABJ,* May 20, 1932; Harold J. Taylor, "Aviation," *ABJ,* June 7, 1932, Nov. 23, 1931; "Darrowville Port Lease Announced," *ABJ,* July 13, 1932; Harold J. Taylor, "Aviation," *ABJ,* Aug. 5, 1932, Feb. 11, 1933; "Learn to Fly," advertisement, *ABJ,* May 12, 1933.

44. "Mid-City Acquired by Three Akronites," *ABJ,* June 28, 1934; "Al Le Fevre Returns as Mid-City Chief," *ABJ,* Dec. 10, 1934; "Mid-City Airport Now in New Hands," *ABJ,* Feb. 6, 1936.

45. "Van Devere Leases Mid-City Airport," *ABJ,* Jan. 25, 1939.

5. Aviation Comes to Kent State University, 1939–41

1. Patricia Strickland, *The Putt-Putt Air Force: The Story of the Civilian Pilot Training Program and the War Training Service, 1939–1944* (Washington, DC: Federal Aviation Administration, 1973), 7–8.

2. Strickland, *Putt-Putt Air Force,* 7–8.

3. "A Brief History of the FAA," *Federal Aviation Administration,* last modified Jan. 4, 2017, https://www.faa.gov/about/history/brief_history; Dominick Pisano, *To Fill the Skies with Pilots: The Civilian Pilot Training Program, 1939–46* (Urbana: Univ. of Illinois Press, 1993), 10.

4. Pisano, *To Fill the Skies with Pilots,* 10–11, 27–28.

5. Pisano, *To Fill the Skies with Pilots,* 3, 5, 8.

6. Karl B. Pauly, *Bricker of Ohio: The Man and His Record* (New York: G. P. Putnam's Sons, 1944), 175; Harold Hanson, "Enrollment Reaches New High," *DKS,* Feb. 10, 1939; KSU Trustees, meeting minutes, Aug. 26, 1939, BOT, KSUSCA.

7. Walker S. Buel, "Ohio Colleges to Train Pilots," *CPD,* Sept. 11, 1939.

8. Strickland, *Putt-Putt Air Force,* 3–10.

9. Strickland, *Putt-Putt Air Force,* 41, 5–9.

10. Pisano, *To Fill the Skies with Pilots,* 145–47.

11. Pisano, *To Fill the Skies with Pilots,* 145–47.

12. "Registrar Filing Applications for Air Course," *DKS,* Sept. 29, 1939; Strickland, *Putt-Putt Air Force,* 10.

13. Accepting women into the program was still contingent upon final word from Washington, but that approval was received by October 2. "Registrar Filing Applications for Air Course"; "Aviation Course Open to University Women," *DKS,* Oct. 2, 1939.

14. "Seventeen Eligible for Flight Course," *DKS,* Oct. 13, 1939; "Three Women Sign for Pilot Course; Runs Total to 47," *DKS,* Oct. 4, 1939. Students identified as being in Kent State's first CPTP class were as follows: Erwin Bailey, Betty Bayes, Arnold Bereit, George Davis, Donald W. Deitz, Rosemary Dempsey, Betty Hayes, Robert Heighton, Clayton Heinlen, Robert C. Hoagland, Henry Russell Kenyon, Ray Lasker, Anton Lysak Jr., Robert J. Matthews, James McCune, Frank Petee Jr., Jack Wardell Petit, William L. Richards, Robert Schwering, Mary Sherriff, Alan Sophrin, Robert Stueve, Fred Swasey, Wilford Terrill, Harold Weil, Wayne Wendell Wilkin, and Mary Jane Wilson.

15. "Seventeen Eligible for Flight Course"; Helen Waterhouse, "Wings over Akron," *ABJ,* Nov. 6, 1939; Pisano, *To Fill the Skies with Pilots,* 145.

16. Helen Waterhouse, "Wings over Akron," *ABJ,* Nov. 6, 1939.

17. Helen Waterhouse, "Wings over Akron," *ABJ,* Nov. 13, 1939; "Good Enough to Teach Student Pilots?," *ABJ,* Nov. 23, 1939.

18. "Potential Pilots Take to the Air," *DKS,* Dec. 1, 1939.

19. Helen Waterhouse, "Wings over Akron," *ABJ,* Dec. 26, 1939.

20. "Private Flying," *Civil Aeronautics Journal* 1, no. 1 (1940): 3–4; Elizabeth Jacobson, "College Girls Eager for Higher Education," *CPD,* Mar. 24, 1940.

21. "Student Pilots Delay Flying," *DKS,* Jan. 8, 1940.

22. "Dietz Solos First in Air Course," *DKS,* Jan. 12, 1940; "Entire Class Starts Flight Work; First Woman Student Soon to Solo," *DKS,* Jan. 19, 1940; Helen Wa-

terhouse, "Wings over Akron," *ABJ,* Jan. 22, 1940, Jan. 29, 1940.

23. Helen Waterhouse, "Wings over Akron," *ABJ,* Feb. 12, 1940, Feb. 26, 1940.

24. "Pilots Complete 700 Flying Hours and Ground Work," *DKS,* Apr. 5, 1940; Richard F. Schwabe, conversation with the author, 1999, Kent State University Airport; "Kent State Students Fly at Akron Airport," *ABJ,* Apr. 8, 1940.

25. "Kent State Students Fly at Akron Airport."

26. Frank Petee, "Reporter Tells How It Feels to Be First KSU Pilot Passed," *DKS,* Apr. 11, 1940.

27. Petee, "Reporter Tells How It Feels to Be First KSU Pilot Passed."

28. Petee, "Reporter Tells How It Feels to Be First KSU Pilot Passed."

29. "Sherriff Is First Ohio Coed Pilot," *DKS,* Apr. 19, 1940.

30. "Lack of Daring Holding Back Women as Aviators," *ABJ,* May 5, 1940; "Akron University, Kent State Students Sprout Wings," *ABJ,* Nov. 30, 1939.

31. "Lack of Daring Holding Back Women as Aviators."

32. "Lack of Daring Holding Back Women as Aviators." Through all of this, Van Wyen, the KSU CPTP flight coordinator, was overseeing activities both on campus and at the airports. The effectiveness of his efforts did not go unnoticed by the CAA, which awarded him a scholarship for his own flight training. The scholarship covered the cost of finishing Stage A of the program, enough to get him to his first solo. The CAA scholarship was given to directors in appreciation for successful integration of the ground and flight courses. "Van Wyen Granted Flight Scholarship," *DKS,* Mar. 13, 1940.

33. Helen Waterhouse, "Wings over Akron," *ABJ,* Apr. 1, 1940, Mar. 11, 1940.

34. Helen Waterhouse, "Wings over Akron," *ABJ,* Mar. 11, 1940, May 20, 1940.

35. Helen Waterhouse, "Wings over Akron," *ABJ,* June 17, 1940, Aug. 26, 1940.

36. Helen Waterhouse, "Wings over Akron," *ABJ,* Apr. 22, 1940. The first woman to join the Kent Flying Club auxiliary was Marguerite Ast, a B. F. Goodrich employee. Helen Waterhouse, "Wings over Akron," *ABJ,* Nov. 3, 1941.

37. Helen Waterhouse, "Wings over Akron," *ABJ,* May 6, 1940; "Birdmen Dine for Graduation," *DKS,* May 9, 1940; "CAA Pilots Finish 1,237 Flying Hours," *DKS,* May 15, 1940; "CAA Still Carries on at Kent State U," *DKS,* Feb. 25, 1941.

38. "Private Flying," *Civil Aeronautics Journal* 1, no. 1 (1940): 3–4; "CAA Pilots Finish 1,237 Flying Hours"; KSU Trustees, meeting minutes, May 17, 1940, BOT, KSUSCA.

39. "Fulton Orders Air Restrictions," *ABJ,* June 4, 1940.

40. Helen Waterhouse, "CAA Opens New Course June 15," *ABJ,* June 7, 1940; "Colleges Offer Flying Courses," *ABJ,* June 12, 1940.

41. Strickland, *Putt-Putt Air Force,* 11; Waterhouse, "CAA Opens New Course."

42. Waterhouse, "CAA Opens New Course"; "Colleges Offer Flying Courses."

43. Helen Waterhouse, "U.S. May Build Barracks Here," *ABJ,* June 17, 1940.

44. Waterhouse, "U.S. May Build Barracks Here."

45. "'California Here We Come,' Future Co-Pilots Wilkins [*sic*] and Schwering," *DKS,* Oct. 1, 1940.

46. "Deadline Is Set for Air School," *ABJ,* June 28, 1940.

47. Helen Waterhouse, "Wings over Akron," *ABJ,* July 29, 1940.

48. "Pilot Falls from Ship as Safety Belt Breaks," *ABJ,* July 29, 1940.

49. "Absolve Van Devere of Blame in Crash," *ABJ,* July 30, 1940; "High Flyers," *ABJ,* Mar. 1, 1981.

50. Pisano, *To Fill the Skies with Pilots,* 83; Frank Petee, "CAA Course Again Offered at Kent State," *DKS,* Sept. 24, 1940.

51. Pisano, *To Fill the Skies with Pilots,* 83.

52. "Air Enlistments to Close Friday," *ABJ,* Sept. 26, 1940.

53. "15 Vacancies Exist in CAA Course," *DKS,* Oct. 4, 1940; "Deadline Nears as Six Enroll for CPT Course," *DKS,* Oct. 9, 1940.

54. "Harter Quizzed on War, Campus ROTC, Invasion," *DKS,* Oct. 16, 1940.

55. Summit County Court of Common Pleas, Sheriff's Deed, no. 394760, July 16, 1936.

56. Helen Waterhouse, "Wings over Akron," *ABJ,* Sept. 30, 1940.

57. KSU Trustees, meeting minutes, Nov. 9, 1940, BOT, KSUSCA; Frank Petee, "Airport Here Materialising; Prexy Hopeful," *DKS,* Dec. 3, 1940.

58. "Industrial College, Airport May Be Established Here; Cost Estimated at $400.000," *DKS,* Nov. 12, 1940.

59. Helen Waterhouse, "Wings over Akron," *ABJ,* Nov. 18, 1940.

60. Helen Waterhouse, "Wings over Akron," *ABJ,* Nov. 18, 1940.

61. Helen Waterhouse, "Akron Becomes Airplane Center Almost Over Night," *ABJ,* Jan. 2, 1941.

62. "Pilot Training Program Offered for Fourth Time," *DKS,* Jan. 14, 1941; "11 Apply for CPTP Prelims,"

DKS, Jan. 17, 1941; "Women to Fly; Senior Pilots Set for Navy," *DKS,* Jan. 21, 1941.

63. "Women to Fly; Senior Pilots Set for Navy"; "CAA Pilot Quota Unfilled; Kent's Flyers Get Wings," *DKS,* Feb. 4, 1941.

64. "Collegiate Newspaper Writers Question Value of CAA Course," *DKS,* Feb. 5, 1941; "CPT Program Budget Is Cut," *DKS,* Feb. 18, 1941; "Freshmen Declared Eligible in Effort to Fill CPT Quota," *DKS,* Feb. 25, 1941.

65. "Flight Course Starts Today for 20 Students," *DKS,* Mar. 6, 1941; "CPT Summer Training Cut off at Kent," *DKS,* Mar. 4, 1941; "Students Pledge Aid to Nation's Defense," *DKS,* Apr. 30, 1941.

66. Helen Waterhouse, "Wings over Akron," *ABJ,* May 5, 1941; Joseph C. Van Devere (son, Rudy Van Devere), phone conversation with the author, Nov. 29, 2016; Helen Waterhouse, "Wings over Akron," *ABJ,* Jan. 12, 1942.

67. As the Depression ran its course, state revenue increased. Ohio had a surplus from gasoline and the annual motor vehicle license tax. From that surplus, $2,517,000 was made available for a building program at the state universities. These funds would be allotted by the Emergency Board to each university for the purchase of its airport. Pauly, *Bricker of Ohio,* 114.

68. George A. Bowman to H. D. Defenbacher, Dec. 30, 1947, Bowman papers, KSUSCA.

69. Ohio State University, Board of Trustees, Record of Proceedings, May 11, 1942, 190, http://kb.osu.edu/dspace/handle/1811/61224; Helen Waterhouse, "Wings over Akron," *ABJ,* May 26, 1941.

70. The CAA's goal was 16,000 young men trained with 10,000 primary and 3,000 secondary students coming from college programs. The remainder would come from noncollege programs, as well as the cross-country and instructor refresher courses. Strickland, *Putt-Putt Air Force,* 56; Helen Waterhouse, "Wings over Akron," *ABJ,* May 30, 1941.

71. Strickland, *Putt-Putt Air Force,* 13.

72. The cross-country course offered over the summer had ten students enrolled. The course consisted of forty hours of flying in a five-seat Stinson airplane. "Summer Courses on CAA Program," *DKS,* May 29, 1941; "Six Student Pilots Earn Advancement," *ABJ,* June 21, 1941.

73. Helen Waterhouse, "Move Civilian Pilot School to Mid-City Due to Traffic Increase at City Port," *ABJ,* Sept. 22, 1941.

74. Waterhouse, "Move Civilian Pilot School to Mid-City."

75. "Co-Pilot Schwering Pays Campus a Visit," *DKS,* Oct. 1, 1941.

76. "Arts College OK's Credit for Airmen," *DKS,* Oct. 10, 1941.

77. Adrian Van Wyen to K. Leebrick, Nov. 24, 1941, Leebrick papers, KSUSCA; Andrew W. Paton, "Aeronautics Training in Higher Education" (draft master's thesis, Kent State University, 1952), II-24, private collection.

78. It was reported a half-dozen Goodyear Aircraft employees were interested in buying a glider and starting a local club. Stow field was mentioned as a perfect site for glider activities since there were no planes flying there. Helen Waterhouse, "Wings over Akron," *ABJ,* Oct. 27, 1941, Dec. 1, 1941.

79. "Andrew Woodburn Paton," Kent State University, Faculty Record, KSUSCA; Ruth Anne Crawford, "Writer, Teacher, Toy Repairer," *DKS,* Nov. 24, 1948; Andrew Paton, Logbook, May 21, 1946 to May 16, 1952, Oct. 5, 1953 to Dec. 31, 1956, private collection; "Paton," Faculty Record, KSUSCA; "KSU's Paton to Train Teachers in Philippines," *DKS,* May 27, 1959; "Andrew Paton—Rich Influence," editorial, *R-C,* Feb. 22, 1964.

6. Wartime at Kent State University, 1941–44

1. Stephen Harding, "First Planes Down at Pearl," *Historynet,* Nov. 4, 2013, http://www.historynet.com/first-planes-down-at-pearl.htm.

2. "Aviation Control Steps Summarized," *Civil Aeronautics Journal* 3, no. 6 (1942): 73, 89.

3. Kenneth Nichols, "Ravenna Gets Double Guard, *ABJ,* Dec. 8, 1941; "CPT Program Ceases until Further Notice," *DKS,* Dec. 10, 1941.

4. Kirby Thorne, "CAA Issues Flight Ban," *New York Times,* Jan. 4, 1942; "Five to Continue CPT Flying Course," *DKS,* Dec. 12, 1941.

5. Patricia Strickland, *The Putt-Putt Air Force: The Story of the Civilian Pilot Training Program and the War Training Service, 1939–1944* (Washington, DC: Federal Aviation Administration, 1973), 8; "CPT Renamed," *Civil Aeronautics Journal* 4, no. 1 (1943): 8.

6. Helen Waterhouse, "Wings over Akron," *ABJ,* Jan. 19, 1942.

7. "Rules Planes Must Use Guarded Ports," *CPD,* Dec. 17, 1941; Helen Waterhouse, "Wings over Akron," *ABJ,* Jan. 19, 1942; "CAA Relaxes Curb on Civilian Pilots," *New York Times,* Jan. 25, 1942; "Weather Forecasts Limited to One Day," *ABJ,* Dec. 17, 1941.

8. "Pilot-Teachers Will Fly Soon," *ABJ,* Dec. 11, 1941.

9. "Flying Ban off Today," *New York Times,* Feb. 15, 1942; "New Regulations for Civilian Aviation Issued," *Civil Aeronautics Journal* 3, no. 4 (1942): 45, 56.

10. "Kent Students Enlist in Service," *DKS*, Jan. 6, 1942; "CAA Spring Pilot Training Aimed at Aiding Army, Navy," *Civil Aeronautics Journal* 3, no. 3 (1942): 29, 43; Andrew W. Paton, "Aeronautics Training in Higher Education" (draft master's thesis, Kent State University, 1952), II-24, private collection; "CPT Increases Quota, Tightens Requirements," *DKS*, Mar. 20, 1942.

11. Paton, "Aeronautics Training," II-15; "CAA Spring Pilot Training."

12. Paton, "Aeronautics Training," II-17, 19.

13. Paton, "Aeronautics Training," II-21–22.

14. Paton, "Aeronautics Training," II-21–22; Dominick Pisano, *To Fill the Skies with Pilots: The Civilian Pilot Training Program, 1939–46* (Urbana: Univ. of Illinois Press, 1993), 97.

15. "C.P.T. Expanded to Serve Needs of Army Air Forces," *Civil Aeronautics Journal* 3, no. 7 (1942): 85.

16. "CPT to Train 10 under New Quota," *DKS*, Mar. 25, 1942; "C.P.T. Expanded to Serve Needs."

17. "26 CPT Pilots Ready to Solo at Airports," *DKS*, July 30, 1942; Van Wyen to Leebrick, memorandum, June 25, 1942, Leebrick papers, KSUSCA; Paton, "Aeronautics Training," II-24; "Thirty CPT Students Get Daily Physical Workouts on Rockwell," *DKS*, July 30, 1942.

18. "Van Wyen Named Dollar-a-Year Man," *DKS*, May 8, 1942.

19. KSU Trustees, meeting minutes, Aug. 25, 1942, BOT, KSUSCA; Paton, "Aeronautics Training," II-20.

20. Paton, "Aeronautics Training," II-18–19.

21. Strickland, *Putt-Putt Air Force*, 19.

22. "CPT Renamed."

23. "CPT-ers Like Kent State; Find It's a 'Great Outfit,'" *DKS*, Jan. 20, 1943; "Weather Halts Airport Work," *DKS*, Jan. 12, 1943.

24. Paton, "Aeronautics Training," II-28–29.

25. "WTS Graduates to Get Pilot Certificates," *Civil Aeronautics Journal* 5, no. 3 (1944): 31.

26. "288 Kent CPT Men Serve in Army, Navy," *DKS*, Feb. 26, 1943; "Kent State Pre-Flight Training," *ABJ*, Feb. 28, 1943; Strickland, *Putt-Putt Air Force*, 12; "2,500 Akron-Trained Pilots Are Flying, Fighting for Uncle Sam," *ABJ*, Mar. 19, 1944; Paton, "Aeronautics Training," II-13.

27. Helen Waterhouse, "Wings over Akron," *ABJ*, Nov. 3, 1941, Mar. 16, 1942.

28. "Fliers in Area Are Grounded," *ABJ*, June 8, 1942.

29. Both Arthur Stanley, Mid-City Airport manager, and Ray Van Devere, Akron Airways manager, charged that the clubs "have been a law unto themselves and members have felt they could do about anything they pleased. They must either have some supervision or remain grounded." William Jarrell, chief inspector of the Cleveland District CAA, traveled to Akron to meet with club members about their grounding and offered that "it will, of course, be temporary and we will have to arrange some sort of supervision." Helen Waterhouse, "Wings over Akron," *ABJ*, Mar. 16, 1942.

30. "Storm Damages Homes, Buildings," *ABJ*, Aug. 10, 1942.

31. Francis H. Dean, *America's Hundred Thousand: The U.S. Production Fighter Aircraft of World War II* (Atglen, PA: Schiffer, 1997), 515–16; Helen Waterhouse, "Private Pilots Voice Complaints," *ABJ*, July 14, 1942.

32. Waterhouse, "Private Pilots Voice Complaints."

33. In May, a site with 231 acres east of Kent on Route 5 near Brady Lake Road was identified. The purchase would involve ten different landowners, include buildings, and necessitate relocating families. The university did not move on this proposal, and in July, the *Kent Stater* referred to the "vain search for a suitable site." L. Wooddell to Leebrick, memorandum, May 21, 1942, Leebrick papers, KSUSCA; "CPT Activity Dull as Plans Lie Dormant," *DKS*, July 2, 1942.

34. Helen Waterhouse, "Wings over Akron," *ABJ*, Aug. 31, 1942; "Study 8 Sites for Airfield at Kent State," *ABJ*, Aug. 1, 1942.

35. KSU Trustees, meeting minutes, Sept. 11, 1942, BOT, KSUSCA; "Ready Stow Field for K.S.U. Students," *ABJ*, Oct. 1, 1942; "University Purchases Stow Flying Field," *DKS*, Oct. 2, 1942. An additional $7,100 was paid as reimbursement for the grading work Van Devere had performed prior to the purchase. KSU Trustees, meeting minutes, Oct. 2, 1942, BOT, KSUSCA.

36. "University Purchases Stow Flying Field"; KSU Trustees, meeting minutes, Oct. 2, 1942, BOT, KSUSCA; "Kent U. Takes over Airport," *ABJ*, Oct. 6, 1942; "Ready Stow Field for K.S.U. Students."

37. "Kent U. Takes over Airport"; "Grade KSU Airport," *DKS*, Nov. 6, 1942.

38. Elsie M. Van Devere and Rudy A. Van Devere to the State of Ohio, Warranty Deed, Oct. 26, 1942, Leebrick papers, KSUSCA; KSU Trustees, meeting minutes, Nov. 17, 1942, BOT, KSUSCA.

39. KSU Trustees, meeting minutes, Jan. 23, 1943, BOT, KSUSCA; "Kent to Have CAA Course," *DKS*, Feb. 17, 1943.

40. KSU Trustees, special meeting minutes, Feb. 11, 1943, BOT, KSUSCA. These units had probably been used during the construction of the Ravenna Army Ammunition Plant.

41. Paton, "Aeronautics Training," III-8–9.

42. William H. Hildebrand, *A Most Noble Enterprise: The Story of Kent State University, 1920–2010* (Kent, OH: Kent State Univ. Press, 2009), 80; Don Webb, "1910–1960: On a University Birthday," *Chestnut Burr* (1960): 19; Kristine Gill, "A Bombing and a Revolution," *DKS*, Feb. 25, 2010; Paton, "Aeronautics Training," III-11.

43. Paton, "Aeronautics Training," III-5–6; "Professors Switch Subjects to Meet Army Demands," *DKS*, Apr. 6, 1943.

44. Paton, "Aeronautics Training," III-6.

45. Paton, "Aeronautics Training," III-7.

46. "Aircrew Will Use University Airport," *DKS*, May 18, 1943; Paton, "Aeronautics Training," III-7; "Riding Along," *DKS*, Apr. 6, 1943; "Future Fliers for Uncle Sam Learn to Get 'Feel of the Air,'" *ABJ*, Apr. 12, 1943.

47. KSU Trustees, special meeting minutes, Jan. 8, 1944, BOT, KSUSCA; Paton, "Aeronautics Training," III-9–10.

48. "Probe Kent Plane Crash Fatal to 4," *CPD*, Mar. 3, 1944.

49. Paton, "Aeronautics Training," III-12; "Cut in Aviation Course Affects 600," *ABJ*, Apr. 1, 1944; Harry H. Ferenz, *Bimonthly Historical Report of the 2586th AAF Base Unit (College Training, Aircrew), Kent State University, Kent, Ohio* (May 1, 1944–June 30, 1944, n.p.); "2,500 Akron-Trained Pilots Are Flying."

50. Glenn Q. Lefler to Arden Allyn, memorandum, Apr. 12, 1944, Bowman papers, KSUSCA.

51. Van Wyen to Earl Crecraft, memorandum, Aug. 16, 1944, Bowman papers, KSUSCA; Raymond M. Clark to Director of Naval Officer Procurement, letter, May 30, 1944, Bowman papers, KSUSCA; "Dr. Dater Receives Promotion," *DKS*, Mar. 13, 1945; Bowman to Van Wyen, letter, Dec. 27, 1944, Bowman papers, KSUSCA; Van Wyen to Bowman, letter, Aug. 9, 1946, Bowman papers, KSUSCA; "Program Friday Will Mark Naval Aviation Anniversary," *Bridgeport (CT) Telegram*, Sept. 25, 1961.

52. Paton, "Aeronautics Training," II-29.

53. Paton, "Aeronautics Training," II-29–30.

54. Paton, "Aeronautics Training," II-30–31.

55. "Akronites Active in Campus Life," *ABJ*, Nov. 7, 1943; KSU Trustees, meeting minutes, Dec. 10, 1943, May 20, 1944, BOT, KSUSCA; "Board of Trustees Presents $2,401,000 Building Program," *DKS*, Apr. 11, 1945.

56. KSU Trustees, meeting minutes, June 8, 1944, BOT, KSUSCA; "Education and Training: History and Timeline," *US Veterans Benefits Administration*, accessed Aug. 28, 2017, https://www.benefits.va.gov/gibill/history.asp.

57. Russellyn "Rusty" Edwards (daughter, Henry Russell Kenyon), phone conversation with the author, Feb. 22, 2017; "Recollections of Lieutenant George Gay, USNR," Naval History and Heritage Command, Sept. 21, 2009, https://www.history.navy.mil/research/library/oral-histories/wwii/battle-of-midway/recollections-of-lieutenant-george-gay.html.

58. Donna Field (niece, Jack Petit), conversation with the author, Jan. 10, 2017; "WW2 War Diaries for VMF-215, Oct. 20, 1943," Fold3, https://www.fold3.com/image/270931780.

59. Helen Waterhouse, "Wings over Akron," *ABJ*, Feb. 26, 1940, July 1, 1940, Oct. 21, 1940.

60. Lowell "Buzz" Starner (son, Lowell Wetzel), conversation with the author, Dec. 4, 2016; Deckmann (Unteroffizier), *Gefechtsbericht, 2./J.G. 27, May 21, 1943* [Battle Report, Group 2/Fighter Squadron 27, May 21, 1943], trans. Michael Everett, private collection; Clifford Hopewell, *Combine 13: A World War II Airman and Prisoner of War in Germany*, 2nd ed. (Austin, TX: Eakin Press, 2000), 8–33.

61. Doug Wilkin (son, Wayne Wilkin), phone conversation with the author, Sept. 26, 2016.

7. Establishing Aviation Education, 1945–49

1. "Board Releases Leebrick," *DKS*, June 1, 1943.

2. James D. Hartshorne, "Northern Ohio Asleep on Airport Expansion," *CPD*, Dec. 13, 1944; Charles B. Donaldson to Bowman, Dec. 27, 1944, Bowman papers, KSUSCA.

3. R. A. Van Devere to Bowman, Aug. 6, 1945, Bowman papers, KSUSCA; Jamie Brian, "A Tribute to the Black Squirrel," *DKS*, Jan. 21, 2015.

4. David Dennis (aviation planner, Ohio Department of Transportation), email message to the author, Apr. 17, 2017.

5. Bowman to R. A. Van Devere, Oct. 16, 1944, Bowman papers, KSUSCA; O. B. Law to Bowman, memorandum, Mar. 23, 1945, Bowman papers, KSUSCA.

6. A year earlier, Van Devere had informed Kent State of a military surplus hangar for sale at Matthews Field in Grady, Arkansas, that could serve the university's own needs, but Bowman passed on the deal. R. A. Van Devere to Bowman, Nov. 24, 1944, Bowman papers, KSUSCA.

7. Van Devere's son Joe remembers it arriving in crates as a surplus government hangar that had been intended for shipment overseas for the war. After it was erected, Joe was sent up to help clean off the rust, only to fall to the ground at least once. R. A. Van Devere to Bowman, Aug. 6, 1945, Bowman papers,

KSUSCA; Bowman to R. A. Van Devere, Sept. 21, 1945, Bowman papers, KSUSCA.

8. Paton to R. A. Van Devere, Nov. 16, 1945, Bowman papers, KSUSCA; R. A. Van Devere to Bowman, Nov. 16, 1945, Bowman papers, KSUSCA.

9. O. B. Law to R. A. Van Devere, Nov. 28, 1945, Bowman papers, KSUSCA; Jean Olsen, "Cadet Returns to Fly at KSU," *DKS,* Jan. 22, 1946; R. A. Van Devere to Bowman, Nov. 22, 1945, Bowman papers, KSUSCA; Helen Waterhouse, "Wings over Akron," *ABJ,* Dec. 28, 1945.

10. Andrew W. Paton, "Aeronautics Training in Higher Education" (draft master's thesis, Kent State University, 1952), III-16, private collection; Bowman to D. F. Barich, memorandum, Oct. 10, 1945, Bowman papers, KSUSCA.

11. Paton to R. A. Van Devere, Nov. 16, 1945, Bowman papers, KSUSCA.

12. Paton, "Aeronautics Training," III-17.

13. "Interest 'High' in Flying; Six Registered Already," *DKS,* Dec. 14, 1945; "University Plans Air Training Program," *DKS,* Oct. 24, 1945; "If You Want to Fly Carry a Full Load," *DKS,* Jan. 25, 1946.

14. Waterhouse, "Wings over Akron," *ABJ,* Jan. 31, 1946; Olsen, "Cadet Returns to Fly at KSU."

15. "High Winds Smash Training Planes," *DKS,* Feb. 15, 1946; Paton to Barich, memorandum, Feb. 18, 1946, Bowman papers, KSUSCA; Barich to Bowman, memorandum, Mar. 11, 1946, Bowman papers, KSUSCA.

16. Paton, "Aeronautics Training," III-22; Waterhouse, "Wings over Akron," *ABJ,* Mar. 27, 1946; "Women Students Enrol [*sic*] in Training Program," *DKS,* Apr. 9, 1946.

17. Waterhouse, "Wings over Akron," *ABJ,* Apr. 17, 1946. Bowling Green State University and Ohio University both had airports closer to their campuses.

18. Paton, "Aeronautics Training," addendum, 2.

19. Paton, "Aeronautics Training," addendum, 3.

20. Paton, "Aeronautics Training," III-19.

21. Paton, "Aeronautics Training," III-20.

22. For the first two quarters of flight training for seventy-five students, Akron Airways was paid $20,095 for its work while Kent State profited $3,643 for the same period. With the impact that the GI Bill was having on flight-training costs, the university had seen costs rise from $2.73 per hour in 1941 to $8.10 per hour solo and $10.80 per hour dual by 1946. Paton to Barich, memorandum, Apr. 10, 1946, Bowman papers, KSUSCA; Bowman to Barich, memorandum, Apr. 22, 1946, Bowman papers, KSUSCA.

23. An instrument course was delayed until financing under the GI Bill could be worked out. "Link Trainer Is Addition to Flight Stock," *DKS,* May 14, 1946.

24. Helen Waterhouse, "Wings over Akron," *ABJ,* Mar. 25, 1947; "Pilots Get New Class Next Term," *DKS,* Feb. 11, 1947.

25. "Buzzers," *DKS,* July 19, 1946.

26. R. A. Van Devere to Larry Wooddell, July 30, 1946, Bowman papers, KSUSCA; KSU Trustees, meeting minutes, Aug. 2, 1946, BOT, KSUSCA.

27. "Kent State Aircraft" was painted over the maintenance hangar, but the name was never officially used. R. A. Van Devere to Bowman, Aug. 12, 1946, Bowman papers, KSUSCA; Bowman to R. A. Van Devere, Sept. 28, 1946, Bowman papers, KSUSCA.

28. "Will Expand Aeronautics," *DKS,* Dec. 11, 1946; Waterhouse, "Wings over Akron," *ABJ,* Mar. 18, 1947.

29. Bowman to Barich, memorandum, Feb. 17, 1947, Bowman papers, KSUSCA.

30. "KSU to Join NIF Clubs," *DKS,* Apr. 23, 1947; Bob Casey, "Local NIFC Branch 'Revs Up' for Wild Blue," *DKS,* May 23, 1947.

31. "IFC Studies 'Cheap' Flying," *DKS,* June 27, 1947; "Flyers Seek New Airplane," *DKS,* July 8, 1947; "Flying Club Has Election," *DKS,* July 11, 1947; "Club to Hear Plane Broker," *DKS,* July 15, 1947; "Flight Club Given 'Q' Hut," *DKS,* Aug. 12, 1947.

32. "Fliers Get New Plane, Officers," *DKS,* Dec. 3, 1947.

33. "Club Seeks Plane," *DKS,* Feb. 27, 1948; "Flying Club Gets New Airplane," *DKS,* Apr. 27, 1948; "KSU Vet, CO-ED 'Fair' Following Plane Crash," *ABJ,* May 22, 1948; "Plane Purchased," *DKS,* Oct. 14, 1948.

34. "Call for Applicants for Air Team," *DKS,* Nov. 9, 1948; "Club Enters Air Meet," *DKS,* Oct. 21, 1948; "Fly Boys Form Team," *DKS,* Oct. 28, 1948.

35. "New Plane for Flying Club," *DKS,* Nov. 11, 1948.

36. "First University Air Meet Scheduled for Thursday," *DKS,* May 17, 1949; "'Into the Wide Blue Yonder,'" *DKS,* May 26, 1949.

37. Bowman to Walter B. Huber, Feb. 10, 1948, Bowman papers, KSUSCA.

38. "Many Shun Flight Course," *DKS,* Aug. 1, 1947; "Seek Improvement in Method of Financing Flight Training," *DKS,* Aug. 8, 1947; "Fall Flight Program Set," *DKS,* Aug. 22, 1947.

39. Paton to Barich, memorandum, Jan. 10, 1948, Bowman papers, KSUSCA.

40. R. A. Van Devere to Bowman, Nov. 18, 1947, Bowman papers, KSUSCA; Bowman to Defenbacher, Dec. 30, 1947, Bowman papers, KSUSCA.

41. John M. Wilcoxon to Bowman, Jan. 7, 1948, Bowman papers, KSUSCA; Bowman to John M.

Wilcoxon, Feb. 12, 1948, Bowman papers, KSUSCA; "University May Buy Airport Buildings," *DKS*, Jan. 21, 1948.

42. "University May Buy Airport Buildings"; Bowman to Frank E. Whittemore, Feb. 12, 1948, Bowman papers, KSUSCA; Bowman to C. E. A. Brown, Feb. 12, 1948, Bowman papers, KSUSCA; KSU Trustees, meeting minutes, May 5, 1948, BOT, KSUSCA.

43. Appraisal of Airport Belonging to Rudy A. Vandevere Located in Township Lot #24, Stow Township, Summit County, Ohio, Mar. 22, 1948, Airport records, KSUSCA.

44. KSU Trustees, meeting minutes, May 4, 1949, Oct. 12, 1949, BOT, KSUSCA.

45. Hank Shapiro, "Aeronautics Course Leaves Student in Air," *DKS*, Aug. 9, 1949.

8. Aviation Takes Hold, 1950–62

1. KSU Trustees, meeting minutes, Feb. 9, 1951, BOT, KSUSCA; Dept. of the Air Force to Bowman, Apr. 20, 1951, Bowman papers, KSUSCA; Kent State University, press release, May 17, 1951, Bowman papers, KSUSCA.

2. Paton to Barich, memorandum, Feb. 23, 1951, Bowman papers, KSUSCA.

3. Bowman to Barich, memorandum, Feb. 24, 1951, Bowman papers, KSUSCA.

4. William G. Workman to Bowman, memorandum, May 21, 1952, Bowman papers, KSUSCA.

5. "Flyers Obtain Lower Prices," *DKS*, Feb. 3, 1950; Don Bean, "Chutes Donned by Vet Pilots," *DKS*, Oct. 22, 1952; "Aviation Courses Offered," *DKS*, May 9, 1952.

6. Chuck Miller, "Flying Club to Begin Sixth Year on Campus," *DKS*, Nov. 14, 1951; Bean, "Chutes Donned by Vet Pilots."

7. Bean, "Chutes Donned by Vet Pilots."

8. R. A. Van Devere to Bowman, May 9, 1952, Bowman papers, KSUSCA.

9. E. W. Tischendorf and Paton to Bowman, memorandum, May 22, 1952, Bowman papers, KSUSCA.

10. Bowman to KSU Trustees, memorandum, Aug. 8, 1952, Bowman papers, KSUSCA; "Discuss Airport," *DKS*, Aug. 21, 1952.

11. Bowman to KSU Trustees, memorandum, Aug. 8, 1952, Bowman papers, KSUSCA.

12. Bowman to R. A. Van Devere, Sept. 2, 1952, Bowman papers, KSUSCA; KSU Trustees, meeting minutes, Oct. 10, 1952, BOT, KSUSCA.

13. Forrest B. Bryant, B. E. Fulton, and Forrest D. Myers, Appraisal of Stow Airport for Kent State University, Feb. 12, 1953, Bowman papers, KSUSCA.

14. Bowman to KSU Board of Trustees, memorandum, Mar. 6, 1953, Bowman papers, KSUSCA; R. A. Van Devere to Bowman, Mar. 18, 1953, Bowman papers, KSUSCA.

15. H. Lee White to Bowman, Mar. 23, 1953, Bowman papers, KSUSCA; KSU Trustees, meeting minutes, Apr. 20, 1953, BOT, KSUSCA.

16. Bowman to R. A. Van Devere, May 5, 1953, Bowman papers, KSUSCA; R. A. Van Devere to KSU Board of Trustees, May 15, 1953, Bowman papers, KSUSCA; KSU Trustees, meeting minutes, Nov. 18, 1953, Jan. 6, 1954, BOT, KSUSCA; Sally Cahur, "Purchase Hangar, Airport Now Complete," *DKS*, Feb. 2, 1954.

17. "'55 Summer School Cut to Five Weeks," *DKS*, Apr. 6, 1954; KSU Trustees, meeting minutes, Oct. 7, 1954, BOT, KSUSCA.

18. Helen Waterhouse, "Wings and Wheels," *ABJ*, Sept. 6, 1953; "GI Bill Provides for Vets to Enroll in Flight Training," *DKS*, Aug. 14, 1952; William H. Hildebrand, *A Most Noble Enterprise: The Story of Kent State University, 1920–2010* (Kent, OH: Kent State Univ. Press, 2009), 84.

19. C. M. Schindler to Bowman, memorandum, Jan. 14, 1953, School of Technology collection, KSUSCA; "KSU Is Only School Offering Lighter Than Airship Training," *DKS*, Jan. 16, 1953; Marvin Katz, "Powered Flights 50th Year," *DKS*, Oct. 20, 1953.

20. The aircraft Kent State received was pieced together from two different Ducks. The wings were painted with the registration NX5502M and came from the third production model. The aircraft fuselage was registration NC5508M, which had first flown on September 22, 1947. That particular airplane saw service at several companies who leased it from Goodyear. Paul Mantz Flying Service first operated the leased Duck in Burbank, California. After a brief stint there, the Duck saw service in Atlanta, Georgia; Jacksonville, Florida; Charleston, South Carolina; and finally Troutdale, Oregon, before being put up for salvage on February 27, 1951. N. H. Hauprich to Paton, Feb. 18, 1959, School of Technology collection, KSUSCA; "'Duck' Added to Aeronautics Dept.," *DKS*, July 12, 1951.

21. Tom Whissen, "Professor Peder Otterson Has Impressive Background," *DKS*, Apr. 21, 1953.

22. Helen Waterhouse, "Wings over Akron," *ABJ*, May 31, 1938; "M. W. Garrett, KSU Professor Emeritus," *ABJ*, July 20, 1974.

23. "Students Give New Life to Old Planes," *DKS*, Feb. 10, 1955.

24. "KSU Aviation Professor Gives Waco Trial Flight," R-C, June 11, 1955.

25. Helen Waterhouse, "Wings and Wheels," *ABJ*, Jan. 31, 1954.

26. "KSU Flying Club Granted Charter," *DKS*, Jan. 26, 1954; "Students Give New Life to Old Planes"; Helen Waterhouse, "Wings-Wheels," *ABJ*, June 20, 1954. The KSU Airport achieved another notable designation in 1955 when it was designated a port of entry. At that point, a pilot could arrive from a foreign country and clear customs at the small airport in Stow. It's unclear how long the KSU Airport retained this designation, but at least for a brief period Kent State owned an international airport. Helen Waterhouse, "KSU Airport Becomes Busy Spot," *ABJ*, Mar. 20, 1955.

27. Paul E. Beck to Bowman, memorandum, July 12, 1955, Bowman papers, KSUSCA; R. C. Van Devere to Beck, Aug. 20, 1955, Bowman papers, KSUSCA; Bowman to Beck, memorandum, Aug. 5, 1955, Bowman papers, KSUSCA.

28. Paton to Bowman, memorandum, Sept. 7, 1955, Bowman papers, KSUSCA; Paton to Berg, memorandum, Nov. 29, 1955, Bowman papers, KSUSCA.

29. Paton to Berg, memorandum, Nov. 29, 1955, Bowman papers, KSUSCA.

30. Berg to Bowman, memorandum, Dec. 5, 1955, Bowman papers, KSUSCA; KSU Trustees, meeting minutes, Dec. 8, 1955, BOT, KSUSCA. Trustee Robert Dix and Paton met with the Ohio Attorney General to explore what options Kent State had to get assistance from the CAA, specifically for funding. The CAA had an Airport Aid Program that gave funds for airport projects, but as Paton learned in this meeting, no state university could participate in the program. It would take more inquiries to find out why that was the case. Paton to Robert Dix, memorandum, Nov. 29, 1955, Bowman papers, KSUSCA.

31. "KSU Boasts Two Active Flight Organizations," *DKS*, Nov. 4, 1954; Paton to Berg, memorandum, Jan. 24, 1956, Bowman papers, KSUSCA; Bruce E. Silcher to Berg, memorandum, Feb. 1, 1956, Bowman papers, KSUSCA.

32. "Airport Gives Way to GM Plant," *Evening Independent* (Massillon, OH), Mar. 8, 1956; Paton to Berg, memorandum, Jan. 24, 1956; Max Lovingood (KSU alumnus), phone conversation with the author, June 14, 2017.

33. Paton to Bowman, memorandum, Mar. 14, 1956, Bowman papers, KSUSCA; C. William O'Neill to Bowman, Apr. 23, 1956, Bowman papers, KSUSCA. KSU Trustees, meeting minutes, Apr. 26, 1956, May 28, 1956, BOT, KSUSCA.

34. Paton to Bowman, memorandum, Jan. 18, 1957, Bowman papers, KSUSCA; Bowman to Paton, memorandum, Jan. 28, 1957, Bowman papers, KSUSCA.

35. "Seek Paved Runway for Airport," *ABJ*, May 16, 1960; Helen Waterhouse, "Private Airports Decrease in Area," *ABJ*, Feb. 21, 1960.

36. John J. Kamerick to Bowman and Robert I. White, memorandum, June 24, 1960, Bowman papers, KSUSCA.

37. Helen Waterhouse, "Wings and Wheels," *ABJ*, Oct. 9, 1960.

38. Paton to Kamerick, memorandum, Apr. 18, 1961, School of Technology collection, KSUSCA.

39. Berg to Bowman, memorandum, July 31, 1961, Bowman papers, KSUSCA; White to Kamerick, memorandum, July 15, 1961, Bowman papers, KSUSCA.

40. Helen Waterhouse, "KSU Gets 9-Passenger Aircraft," *ABJ*, Apr. 2, 1961.

41. Paton to Kamerick, memorandum, Feb. 1, 1962, Bowman papers, KSUSCA; White to Kamerick, memorandum, Mar. 6, 1962, Bowman papers, KSUSCA.

42. Bowman to Paton, memorandum, Sept. 13, 1955, Bowman papers, KSUSCA.

9. Total University Control at Last, 1963–71

1. Michael R. Mack, Leonard's son, flew with his dad and Robert Dix, KSU trustee and *Record-Courier* publisher. He remembers Dix saying he hoped he would live long enough to see a paved runway. He did. The runway was paved in 1965, and Dix died in 1996. Dix owned the airplane used in the *Ripcord* TV series, a Cessna 210. Michael R. Mack (son, Leonard Mack), discussion with the author, July 13, 2017; J. W. Bunn to White, memorandum, July 29, 1963, White papers, KSUSCA.

2. Bunn to White, memorandum, Sept. 9, 1963, White papers, KSUSCA.

3. White to KSU Board of Trustees, memorandum, Oct. 8, 1963, White papers, KSUSCA. The title of the head of the flight program varied; from this point forward, the title "director" will be used.

4. Bill Bierman, "Up in the Air KSU Airport Is Valuable, But to Whom, White Asks," *DKS*, Oct. 15, 1963; Robert E. Hill to University Airport Committee, Oct. 31, 1963, White papers, KSUSCA.

5. Andrew W. Paton, "Brief History Kent State Airport," Oct. 7, 1963, White papers, KSUSCA; Andrew W. Paton, "Pattern of University Use of the

University Airport," Dec. 2, 1963, White papers, KSUSCA.

6. Paton, "Brief History Kent State Airport"; Paton, "Pattern of University Use of the University Airport."

7. Paton, "Brief History Kent State Airport"; Paton, "Pattern of University Use of the University Airport."

8. KSU Airport Special Committee, meeting minutes, Nov. 9, 1963, White papers, KSUSCA.

9. KSU Airport Special Committee, meeting minutes, Dec. 11, 1963, White papers, KSUSCA.

10. "Andrew Paton Dies; Saturday Rites Set," *DKS,* Feb. 21, 1964; "KSU Airport to Be Renamed in Honor of Andrew W. Paton," *DKS,* July 14, 1964.

11. White to Marshall W. Garrett, memorandum, Dec. 27, 1963, White papers, KSUSCA; Helen Waterhouse, "Would Executive Planes Aid KSU Port?," *ABJ,* Dec. 15, 1963; KSU Airport Special Committee, meeting minutes, Mar. 25, 1964, White papers, KSUSCA.

12. KSU Airport Special Committee, meeting minutes, Apr. 8, 1964, White papers, KSUSCA; KSU Airport Special Committee, Future Plans for the Educational Aspects of the University Airport at Kent State, Mar. 16, 1964, White papers, KSUSCA.

13. Delmar W. Olson to Charles C. Tillinghast Jr., president, TWA, Mar. 9, 1964, White papers, KSUSCA; School of Technology, KSUSCA; Olson to Hill, Apr. 21, 1964, White papers, KSUSCA.

14. KSU Airport Special Committee, "Improvement of the KSU Airport: Federal Aid Advantages and Disadvantages," Mar. 18, 1964, White papers, KSUSCA.

15. KSU Airport Special Committee, "Kent State University Airport: A Report," June 24, 1964, White papers, KSUSCA.

16. KSU Airport Special Committee, "Kent State University Airport: A Report."

17. KSU Airport Special Committee, "Kent State University Airport: A Report."

18. KSU Trustees, meeting minutes, July 9, 1964, BOT, KSUSCA; Olson to White, memorandum, June 2, 1965, White papers, KSUSCA; William A. Paton, in communication with the author, Oct. 27, 2017. It would take some time before the name would become standardized. By 1966, Andrew W. Paton Field was painted on the hangar, while the sign on the street side was Andrew W. Paton Airport. In the 1980s, the street sign was replaced, showing Kent State Airport without Paton's name. The next airport sign, installed in 2012, identified the Kent State University Airport and brought back the name of Andrew W. Paton Field, matching the name on the hangar. Paton Field will be the convention used through the remainder of this book.

19. KSU Trustees, Belcher Real Estate Option, July 9, 1964, BOT, KSUSCA; KSU Trustees, meeting minutes, Sept. 17, 1964, BOT, KSUSCA.

20. "Urge Akron Buy, Move KSU Airport," *ABJ,* Sept. 23, 1964; "Urge Stow to Endorse KSU Airport Switch," *ABJ,* Sept. 25, 1964; Al Stevenson, "Will Fight on Against Expansion of KSU Airport," *ABJ,* Oct. 6, 1964; J. W. Bunn to White, memorandum, Oct. 13, 1964, White papers, KSUSCA.

21. D. E. Dominic to White, Sept. 25, 1964, White papers, KSUSCA; White to D. E. Dominic, Mar. 20, 1965, White papers, KSUSCA. A new high school with athletic fields would later be built on east Graham Road in Stow, opening in Sept. 1987.

22. KSU Trustees, meeting minutes, July 12, 1965, BOT, KSUSCA; Mack, discussion with the author, July 13, 2017.

23. Olson to Perlmutter, memorandum, May 4, 1965, White papers, KSUSCA; Stan Calhoun, "Kiwi Bird 'Nests' at KSU Airport," *DKS,* Feb. 11, 1966.

24. P. A. Otterson to Olson, memorandum, May 25, 1965, White papers, KSUSCA; Peder Adolph Otterson, "An Evaluation of the Aerospace Technology Program at Kent State University" (PhD diss., Ohio State University, 1968).

25. Marshall W. Garrett and Manley J. Morrison to White, memorandum, Sept. 3, 1965, White papers, KSUSCA; Marshall W. Garrett and Manley J. Morrison, Kent State University Airport, "Master Plan for Andrew W. Paton Airport, Sept. 3, 1965," White papers, KSUSCA.

26. Garrett and Morrison, "Master Plan for Andrew W. Paton Airport."

27. KSU Trustees, meeting minutes, Sept. 16, 1965, BOT, KSUSCA.

28. David Mohan, "Analysis, Philosophical Perspective, and Recommendations concerning Aviation Education at Kent State University," Dec. 12, 1989, private collection; Olson to Perlmutter, memorandum, Oct. 12, 1965, White papers, KSUSCA.

29. Olson, "The Kent State University A. W. Paton Airport Operation: A Proposal," Apr. 8, 1966, White papers, KSUSCA.

30. Olson to Kamerick, memorandum, Apr. 8, 1966, White papers, KSUSCA.

31. Olson, "The Kent State University A. W. Paton Airport Operation."

32. Olson, "The Kent State University A. W. Paton Airport Operation."

33. In August, the Airport Committee recommended that negotiations be started with Van Devere to acquire airplanes from those he needed to

divest. Emil Berg to White, memorandum, July 25, 1966, White papers, KSUSCA; Airport Committee, meeting minutes, Aug. 25, 1966, White papers, KSUSCA; Olson to Kamerick, memorandum, July 23, 1966, White papers, KSUSCA.

34. Olson to Scott Henninger, Sept. 16, 1966, White papers, KSUSCA.

35. Richard Frank Schwabe. *Student Pilot Rating Book,* 1941, extracted pages, private collection; Francis Pike, *Hirohito's War: The Pacific War, 1941–1945* (New York: Bloomsbury, 2015), 336; Jennifer Schwabe Ramsey, email message to the author, Feb. 2, 2018; Jean Schwabe (wife, Richard Schwabe), discussion with the author, July 7, 2016.

36. Olson to Kamerick, memorandum, Sept. 23, 1966, White papers, KSUSCA.

37. Otterson, "An Evaluation of the Aerospace Technology Program," 16; Eugene G. Ripple, "History of Aerospace Technology at Kent State University, 1941–1976" (draft master's thesis, Municipal University of Akron, Mar. 1976?), 33.

38. KSU Trustees, meeting minutes, Oct. 20, 1966, BOT, KSUSCA.

39. Thrity Umrigar, "Freedom: It Rings Loudly for Some," *ABJ,* July 4, 1993.

40. Umrigar, "Freedom"; David Dix, "Along the Way: From Stow, Ohio, to the Smithsonian Air and Space Museum," *Twinsburg (OH) Bulletin,* May 24, 2012; "Albert C. Beckwith," *Courier-News* (Bridgewater, CT), Oct. 8, 2014.

41. Carole Edwards and Harry Liggett, "Mysterious Plane Crash Kills Two," *ABJ,* Oct. 22, 1966.

42. William VonGunten (KSU alumnus), discussion with the author, Apr. 21, 2017; Bunn to Airport Committee, memorandum, Dec. 2, 1966, White papers, KSUSCA; White to J. W. Bunn, memorandum, Jan. 26, 1967, White papers, KSUSCA.

43. Smith's primary responsibility would be instructing aerospace technology students, while Schwabe's would first be for the ROTC cadets and then for aerospace technology students. Olson to John A. Flower, memorandum, Jan. 16, 1967, White papers, KSUSCA; "KSU Purchases Two Airplanes," *DKS,* Feb. 16, 1967; Flower to John J. Kamerick, memorandum, May 26, 1967, White papers, KSUSCA.

44. Aerospace Technology Division, "Airport Facilities—Preliminary Requirements," Feb. 15, 1967, White papers, KSUSCA; Flower to Bunn, memorandum, Feb. 16, 1967, White papers, KSUSCA.

45. Ripple, "History of Aerospace Technology," 33; "Ripple, Eugene G.," Service Record, Kent State University, KSUSCA; "Dr. Eugene George Ripple," obituary, *ABJ,* Mar. 5, 2015; Richard McBane, "Love for Flying Runs in Family," *ABJ,* Mar. 13, 1995; Kamerick to White, memorandum, May 29, 1967, White papers, KSUSCA; White to Kamerick, memorandum, June 1, 1967, White papers, KSUSCA.

46. "Bus Service Begins Feb. 13," *DKS,* Jan. 31, 1967; Campus Bus Service, Management Operations Report, Aug. 1967, private collection.

47. "Bus Times Rescheduled," *DKS,* Feb. 8, 1968; Ted Orris, letter to the editor, *DKS,* Feb. 6, 1981; Rick Lackner, "Campus Bus Service Changes Routes," *DKS,* Aug. 31, 1993.

48. Flower to White, memorandum, July 2, 1969, White papers, KSUSCA; "KSU Pilot Mack is Dead at 52," *DKS,* July 10, 1969.

49. KSU School of Technology, (Preliminary) Basic Data Sheet-IAT 494, July 18, 1969, White papers, KSUSCA; Otterson to White, memorandum, July 23, 1969, White papers, KSUSCA; R. E. Dunn to White, memorandum, July 30, 1969. White papers, KSUSCA; White to Otterson, memorandum, Aug. 4, 1969, White papers, KSUSCA.

50. "KSU Gets Airplane as Gift," *R-C,* Feb. 5, 1970; Ripple, "History of Aerospace Technology," 34.

51. Aerospace Technology, Program Description, Feb. 20, 1970, School of Technology Collection, KSUSCA.

52. Jerry M. Lewis and Thomas R. Hensley, "The May 4 Shootings at Kent State University: The Search for Historical Accuracy," *The Ohio Council for the Social Studies Review* 34, no. 1 (1998), accessed Sept. 24, 2017, https://www.kent.edu/may-4-historical-accuracy; Karen S. Lillis, "Prologue to Tragedy," *DKS,* May 4, 1995; Ohio National Guard HHC 1–145, Daily Staff Journal, May 3, 1970, Charles A. Thomas papers, May 4 Collection, KSUSCA.

53. "Kent State Police Log Tells a Dramatic Story," *R-C,* May 4, 1970; "Vandals Damage 6 Planes at Portage County Airport," *R-C,* May 4, 1970; Jean Schwabe (wife, Richard Schwabe), discussion with the author, July 25, 2016.

54. Ohio National Guard HHC 1–145, Daily Staff Journal, May 4, 1970, Charles A. Thomas papers, May 4 Collection, KSUSCA.

55. Howard Means, *67 Shots: Kent State and the End of American Innocence* (Boston: Da Capo, 2016), 172; Ohio National Guard HHC 1–145, Daily Staff Journal, May 5?, 1970, Charles A. Thomas papers, May 4 Collection, KSUSCA.

56. David N. Hawk to William Herr, July 10, 1970, White papers, KSUSCA; White to Hoagland, memorandum, June 15, 1970, White papers, KSUSCA;

KSU Trustees, meeting minutes, Aug. 20, 1970, BOT, KSUSCA.

57. KSU Aerospace Technology Division, *Alumnus News,* June 1969, School of Technology Collection, KSUSCA; Tom Craig, "Space Limits Aerospace," *DKS,* Dec. 4, 1970. While plans for a new airport building did not go forward, major construction was under way on campus for a twelve-story library building. To build a book collection to go along with the building, White authorized the purchase of 250,000 rare books from Gilman's bookstore in Crompond, New York. The acquisition of the 325-ton collection was delivered by seven semitrailer trucks to the airport's hangar, which served as temporary storage for the 7,100 boxes. Dean Keller (former head of KSU Special Collections and Archives), discussion with the author, Jul. 24, 2018; "Library Adds 250,000 Books!," *DKS,* Aug. 22, 1968.

58. Eric E. Johnson, "KSU Offers Flight Training Program," *DKS,* Apr. 22, 1971.

59. "Aerospace Courses to Focus on Occupational Areas," *DKS,* May 25, 1971.

10. Seeking Financial Stability, 1971–82

1. James Ricci, "Olds Begins Work on His Dreams for KSU," *ABJ,* Sept. 16, 1971; William H. Hildebrand, *A Most Noble Enterprise: The Story of Kent State University, 1920–2010* (Kent, OH: Kent State Univ. Press, 2009), 178.

2. Tom J. Foley, letter to the editor, *DKS,* Apr. 13, 1972; H. L. Harrison, letter to the editor, *DKS,* May 25, 1972; Jim Morris and Dana J. Holmes to Glenn A. Olds, letter, Apr. 5, 1972, Glenn A. Olds papers.

3. School of Technology, Aerospace-Airport Budget Brief, rev. June 14, 1972, Olds papers; Charles W. Keith to Ralph Hetzel, memorandum, June 15, 1972, Olds papers; John A. Flower to Ronald W. Roskens, memorandum, June 20, 1972, Olds papers.

4. The cost for flight training was $600 per quarter. Rental costs were $15 per hour for a Cessna 150 and $17 per hour for a Cessna 172. Paul Bossman, "Fly This Summer through KSU's Flight Training Program," *DKS,* May 30, 1972; "Aircraft Available for Rent," *DKS,* Dec. 6, 1972; "Airport Gets New Aircraft," *DKS,* Oct. 17, 1972; Elizabeth Sappenfield, "Airport Offers Services," *DKS,* Dec. 7, 1972.

5. Otterson to Olds, "Airport—Position Report, Feb. 16, 1973," Olds papers; Hetzel to Hall, memorandum, Feb. 20, 1973, Olds papers.

6. Aerospace Technology Division, "Outline for the Establishment of a School of Aviation and Transportation," Apr. 1973, Olds papers.

7. Aerospace Technology Division, "Outline for the Establishment of a School of Aviation and Transportation."

8. Hetzel to Hall, memorandum, Apr. 19, 1973, Olds papers.

9. Hall to Olds, memorandum, June 8, 1973, Olds papers; Daniel L. Newcomb to Fay R. Biles, memorandum, June 13, 1973, Olds papers; Otterson to Olds, memorandum, June 11, 1973, Olds papers.

10. "Hoagland New Traffic Director," *DKS,* Aug. 3, 1972; "Airport Gets New Aircraft"; William H. Wilgus to university administrators, memorandum, Feb. 5, 1973, Olds papers.

11. James Lawson, "Stow Land Development Hinges on Airport Plans," *ABJ,* Mar. 11, 1974.

12. Lois LaDow, "Chamber Endorses Airport Goal," *Stow (OH) Sentinel,* Apr. 19, 1973; Fran Arman, "Airport 'Master Plan' Still 'Up in the Clouds,'" *DKS,* Apr. 27, 1973.

13. Bruce Larrick, "Airport Backers Wonder: 'Where Do We Stand,'" *ABJ,* Apr. 2, 1973; John A. Doyle to Glenn A. Olds, letter, Apr. 13, 1973, Olds papers; Glenn A. Olds to Dana J. Holmes, letter, Apr. 16, 1973, Olds papers; Glenn A. Olds to Thomas F. Troyer, letter, May 8, 1973, Olds papers.

14. Hetzel to Hall, memorandum, June 16, 1972, Olds papers; Olds to Flower, memorandum, July 17, 1972, Olds papers.

15. Otterson to Olds, memorandum, Sept. 28, 1972, Olds papers.

16. Otterson to Newcomb, memorandum, Dec. 11, 1972, Olds papers.

17. Otterson to Hetzel, memorandum, Feb. 6, 1973, Olds papers.

18. "Dome 'Strung Up,'" *DKS,* May 19, 1972; Aerospace Technology Division, "Aerospace Technology Facility Proposal," n.d., Olds papers; Otterson to Olds, Sept. 27, 1972, Olds papers; Otterson to Biles, memorandum, Sept. 13, 1972, Olds papers.

19. Robert S. Ross to Keith, letter, Feb. 27, 1973, Olds papers; Herbert Q. Bair to Glenn A. Olds, letter, May 23, 1973, Olds papers.

20. Biles to Olds, memorandum, Apr. 11, 1973, Olds papers.

21. Newcomb to Olds, memorandum, May 21, 1973, Olds papers; Arnold Dengley, "Portage Airport Invites KSU to Use Its Facilities, *DKS,* May 10, 1974.

22. Olds to John Goodwin, letter, Sept. 28, 1973, Olds papers; Arnold Dengler, "Federal Funds Requested," *DKS,* May 8, 1974; KSU Trustees, meeting minutes, July 17, 1975, BOT, KSUSCA.

23. Kent State University and TransPlan, Inc., *Master Plan for Kent State University Airport: An-*

drew W. Paton Field (New York: TransPlan, 1977), 1:1; Walter G. Bruska to Olds and Board of Trustees, memorandum, Jan. 26, 1977, BOT, KSUSCA; "$4 Million Expansion of KSU Airport Urged," *CPD,* Nov. 2, 1976.

24. Ken Hiller, "Plan for Airport Improvement Opposed by Stow Residents," *DKS,* Oct. 19, 1977.

25. "Crash Rocks Joe Walsh," *ABJ,* Mar. 15, 1975; "NTSB Identification: CHI75FEI39," National Transportation Safety Board, Mar. 14, 1975, https://www.ntsb.gov/_layouts/ntsb.aviation/brief.aspx?ev_id=47764&key=0.

26. "Stow Panel Backs Airport Work," *ABJ,* Jan. 11, 1978; "Airport Expansion Wins Round One," *Stow (OH) Sentry,* Jan. 18, 1978; Mark Barnette, "Airport Receives Improvements Worth $4 Million," *DKS,* Sept. 19, 1980; Ross Sneyd, "FAA to Aid Paving Projects at KSU Airport," *DKS,* July 28, 1982.

27. Jack Williams, "$98,644 for Trainees' Planes," *DKS,* Oct. 3, 1973; Otterson to Faculty Senate Executive Committee, memorandum, Dec. 11, 1973, Olds papers.

28. Eugene G. Ripple, "History of Aerospace Technology at Kent State University, 1941–1976" (draft master's thesis, Municipal University of Akron, Mar. 1976?), 40; Commercial Operator's Agreement, 1979–80, private collection.

29. "Booster Club Promotes Fund Raising," *DKS,* May 21, 1974; Olds to Robert M. White, AFROTC Commander, letter, Oct. 24, 1974, Olds papers.

30. Kent State University, *Financial Report, June 30, 1976 (Kent, OH: Kent State University),* Schedule D-8, BOT, KSUSCA; Jeannine Guttman, "Olds, VP's Make Move; More Building Planned," *DKS,* Sept. 23, 1976.

31. "Olds Sets Plans for Flight Course," *DKS,* June 1, 1973; Joan Kobosky, "$1 Million Cut Seen in 1977–78 Budget," *DKS,* Feb. 17, 1977; Joan Kobosky and Paul Shane, "Farewell to KSU," *DKS,* June 3, 1977; "Olds Honored by Legislature, Staff," *DKS,* Mar. 1, 1977.

32. Hildebrand, *Most Noble Enterprise,* 217–18; Beth Belton, "Three VP Positions Cut in Reorganization," *DKS,* Jan. 31, 1978.

33. Cari Orris and Jeff Gallatin, "U-School Just Too Elite, Golding Says," *DKS,* Nov. 19, 1981.

34. P. A. Otterson, "Campus News," *Technology Division Newsletter, 1978–79,* Jan. 1979, 2; Mary Lou Sneyd, "More Flight Students up in Air," *DKS,* Sept. 7, 1979; Dave Wooldridge, "Airport Improvement Funds Being Held," *DKS,* Aug. 6, 1981.

35. Kim Nero, "Simulators Keep Pilots Flying," *DKS,* Feb. 24, 1982; Tim Roske, "Flight Program Violates FAA Rules," *DKS,* Nov. 7, 1986; Tom Grossman (former KSU flight-training director), phone conversation with the author, Oct. 30, 2017.

36. Wooldridge, "Airport Improvement Funds Being Held."

37. Wooldridge, "Airport Improvement Funds Being Held"; Grossman, phone conversation with the author, Oct. 30, 2017. As of 2017, the trailer is still called the FAA trailer, although testing was moved to campus in 1996.

38. KSU Trustees, meeting minutes, Dec. 11, 1980, BOT, KSUSCA; Ted Orris (former KSU assistant chief flight instructor), phone conversation with the author, Oct. 19, 2017.

39. "Flying Club Seeks Members," *DKS,* Oct. 26, 1972; Shelly Giroski, "New Cessna Training Plane Assists Flying Club Needs," *DKS,* May 8, 1979.

40. Paul Bossman, "Flying Club to Take Part in Tourney," *DKS,* Apr. 14, 1972; "Flying Club to Compete in National Championships," *DKS,* May 8, 1973; Chuck Poliafico, "University Flying Team Places 11th at National Meet," *DKS,* June 19, 1980.

41. "Air Expo Plans Get off Ground," *DKS,* Apr. 20, 1973; Hildebrand, *Most Noble Enterprise,* 219.

42. Arnold Dengler, "Air Expo to Promote KSU Aviation Activities," *DKS,* May 10, 1974; "Airshow Presents Jets, Beauty," *DKS,* Apr. 6, 1976; Sue Christopher, "Air Expo!," *DKS,* May 5, 1977.

43. Nathaniel Madison, "Aerobatic-Acrobatics Highlight Expo Show," *DKS,* May 19, 1978; Shelly Giroski, "'Falling Angels' Open Air Show," *DKS,* May 8, 1979; Chuck Poliafico, "Death-Defying Aerial Stunts Will Highlight Air Expo '80," *DKS,* May 5, 1980; "Air Show to be Held Saturday," *DKS,* May 8, 1981; Mark A. Williamson, "Airport Show Features Flying Display, Aircraft Rides," *DKS,* Apr. 23, 1982; "Razzle Dazzle in the Air," *DKS,* May 15, 1979. Schwabe's retirement from Kent State had left a void as he had been teaching all the ground schools, in addition to flight training and testing at the airport. Grossman, phone conversation with the author, Oct. 30, 2017.

44. In 1982, sponsors obtained a $3,000 liability insurance policy from Lloyd's of London. Orris, phone conversation with the author, Oct. 19, 2017; Giroski, "'Falling Angels' Open Air Show."

11. First Improvements, Then a Threat, 1982–91

1. Michael Schwartz, "Historical Note to the Mission Statement," May 8, 1985, BOT, KSUSCA; Frank Badillo, "Streamline KSU, Schwartz Says," *DKS,* Oct. 21, 1982; "KSU's Mission Statement Adopted; Defines Emphasis," *DKS,* Sept. 13, 1983.

2. Ross Sneyd, "FAA to Aid Paving Projects at KSU Airport," *DKS,* July 28, 1982.

3. KSU Trustees, meeting minutes, Sept. 30, 1982, BOT, KSUSCA.

4. KSU Trustees, meeting minutes, Feb. 24, 1983, BOT, KSUSCA.

5. KSU Trustees, meeting minutes, May 26, 1983, BOT, KSUSCA; KSU Trustees, meeting minutes, Sept. 9, 1983, BOT, KSUSCA.

6. KSU Trustees, meeting minutes, May 29, 1985, BOT, KSUSCA.

7. KSU Trustees, meeting minutes, Mar. 11, 1987, BOT, KSUSCA.

8. Norman Leigh, "Pilot Safety Key Factor in Tree Clearing," *DKS,* June 22, 1983.

9. Tom Grossman (former KSU flight-training director), phone conversation with the author, Oct. 30, 2017.

10. Molly Callahan, "Airport Expansion Taking Off," *DKS,* Dec. 12, 1986; KSU Trustees, meeting minutes, Sept. 23, 1987, BOT, KSUSCA.

11. KSU Trustees, Committee on Business Affairs and Institutional Advancement, Summary Report of Meeting, June 1, 1987, BOT, KSUSCA. For several weeks during construction of the runway extension, some flight operations were moved to Akron-Fulton Municipal Airport. Ripple arranged for KSU airplanes to be stored in the Goodyear Airdock. Grossman, phone conversation with the author, Oct. 30, 2017. The airport was renamed "Akron-Fulton Municipal Airport" in 1978 to honor both "Shorty" Fulton and his son, "Bud," who between them had managed the facility from its beginning through 1977. William Hershey, "Abortion Vote Hasn't Halted Council Feud," *ABJ,* Mar. 1, 1978; Grossman, phone conversation with the author, Oct. 30, 2017.

12. KSU Trustees, Committee on Business Affairs and Institutional Advancement, Summary Report of Meeting, Nov. 22, 1983, BOT, KSUSCA; Grossman, phone conversation with the author, Oct. 30, 2017.

13. "KSU Airport Receives State Grant," *DKS,* Apr. 14, 1989.

14. Ripple to customers and friends, letter, Mar. 1, 1984, private collection; Albert C. Beckwith to Ripple, letter, Apr. 4, 1984, private collection; Beckwith to Ripple, letter, Dec. 21, 1984, private collection; Ripple to customer, letter, May 10, 1985, private collection.

15. There were maintenance issues with the existing T-hangars, which were now twenty years old. In November 1985, Beckwith reported leaking roofs in four of the hangars he was renting. In 1989, the situation had not improved, and he expressed concern over T-hangar doors, security locks, and limit switches. Tim Roske, "Space Problem Leaves Airport Hanging," *DKS,* Oct. 17, 1986; Beckwith to Ripple, letter, Nov. 21, 1985, private collection; Beckwith to Ripple, letter, Mar. 14, 1989, private collection.

16. Christine Klecic, "KSU Airport Hangar Loses Roof to Winter," *DKS,* Jan. 27, 1984. Concerned Alumni to Schwartz, letter, Aug. 2, 1984, Schwartz papers, KSUSCA; Hortense McGehee to Schwartz, letter, Aug. 27, 1984, Schwartz papers, KSUSCA.

17. David Mohan to Terry Roark, memorandum, Aug. 30, 1984, Schwartz papers, KSUSCA.

18. Steve Wright, "Airport's Additions Made for Safety's Sake," *DKS,* Feb. 14, 1985.

19. Mike Flanagan (KSU flight program alumnus), phone conversation with the author, Oct. 12, 2017. The new airplanes received these Kent fleet numbers—Cessna 152s: Kent 1–8 and Kent 11–12 (from Wichita); Kent 13–15 (from Minot); Kent 16 (possible lease from Akron-Fulton operator); Cutlasses: Kent 20–23; Cessna 172s: Kent 24–28. Grossman, phone conversation with the author, Oct. 30, 2017. The original flight-slot ledger allowed for only eight entries per page. The decision was made to number the fleet 1–8, 11–18, 21–28.

20. "Flight Team is Region 3 Titlist," *DKS,* Dec. 9, 1982; Sallyann Mogford, "Flight Team Third at Regionals," *DKS,* Oct. 8, 1985.

21. Ted Orris, assistant chief flight instructor, commented that at times there were up to sixty people in the trailer. "Airport Completing Land Expansion," *DKS,* Aug. 25, 1986; Tim Roske, "Flight Program Violates FAA Rules," *DKS,* Nov. 7, 1986.

22. "Student in Serious Condition," *DKS,* Apr. 9, 1986; "NTSB Identification LAX86FA167," National Transportation Safety Board, Aug. 11, 1988, https://www.ntsb.gov/_layouts/ntsb.aviation/brief.aspx?ev_id=20001213X33294; Susanna Brandon, "KSU Plane Crash Victim Sues Maintenance Company," *DKS,* Apr. 7, 1988; Craig Paeth, "KSU paid $909,000 in Crash Settlement," *DKS,* Nov. 3, 1989; Flanagan, phone conversation with the author, Oct. 12, 2017.

23. Mary Vanac, "Ex-Pilot Voices Residents' Concerns about Airport," *DKS,* Jan. 19, 1989; Lisa Madel, "Stow Ponders Airport," *ABJ,* Sept. 28, 1989; "City and Region," *ABJ,* May 25, 1989; Beckwith to airport supporters, letter, June 22, 1989, private collection; Beckwith to airport neighbors, letter, July 1989, private collection.

24. The consequence of raising the flight pattern altitude was that it also changed the type of airspace that the traffic pattern operated in. Previously, only one mile of visibility was required due to it being in Class G airspace. The new traffic pattern was in Class E airspace, requiring three miles of visibil-

ity. Flanagan, phone conversation with the author, Oct. 12, 2017; "Meetings of Note," *ABJ,* Oct. 10, 1989; Craig Webb, "KSU Airport Acts for Stow," *DKS,* Oct. 4, 1989; Lisa Madel, "Stow Orders Jet Ban," *ABJ,* Oct. 13, 1989; Beckwith to Stow City Council members, letter, Oct. 10, 1989, private collection.

25. Randolph Reynolds, "Airport Neighbors: KSU Misled Us," *DKS,* Mar. 24, 1989; Craig Webb, "Trustees Approve New Faculty Contract," *DKS,* Sept. 14, 1989.

26. Lisa Madel, "Stow Council May Use Zoning in Airport Fight," *ABJ,* Dec. 6, 1989; "Meetings of Note," *ABJ,* Dec. 15, 1989; Joe Cowley and Percy Allen, "Stow City Council Passes Ordinance Giving the City Greater Control of Airport," *DKS,* Feb. 9, 1990.

27. David Poluga (KSU airport manager), email message to the author, June 23, 2017; Peter Geiger, "Small Airports Called Big Attractions," *ABJ,* Aug. 8, 1988; Bill Bush, "Recalling Biplanes and Bygone Days," *DKS,* Jan. 29, 1988.

28. Steve Berend, "Students Say Flight Program Should Taxi Improvements," *DKS,* May 6, 1988.

29. Moore to Schwartz, memorandum, Jan. 19, 1988, Schwartz papers, KSUSCA.

30. C. Hall Jones and George Stragisher to Schwartz, letter, Nov. 10, 1988, Schwartz papers, KSUSCA; Mohan to Thomas J. Barber, memorandum, May 5, 1989, Albert C. Beckwith papers, private collection.

31. Cheryl A. Casper to Moore, memorandum, June 15, 1989, Schwartz papers, KSUSCA.

32. Moore to Barber, memorandum, Oct. 30, 1989, Schwartz papers, KSUSCA; Cristal Williams Walker, "KSU Ends Its Pilot Training," *ABJ,* Nov. 3, 1989; Trish Hooper, "Stow Officials Surprised by Decision," *DKS,* Nov. 3, 1989.

33. Cristal Williams Walker, "Heated Meeting at KSU over Flight Program Fate," *ABJ,* Nov. 8, 1989; OCAE Information Bulletin, n.d., Scott Koeppl papers, private collection; Rebecca Buckingham and Maya Eckstein, "End of Flight Program Not Accepted by All," *DKS,* Nov. 8, 1989; Steve Di Penna, "KSU Instructors Discuss Ways to Save Program," *DKS,* Nov. 8, 1989; Scott Koeppl, ('89 alumnus), email message to the author, June 5, 2017.

34. Maya Eckstein, "Airport Operated at a Deficit," *DKS,* Nov. 9, 1989; KSU Trustees, meeting minutes, Nov. 15, 1989, BOT, KSUSCA.

35. David Mohan, "Analysis, Philosophical Perspective, and Recommendations Concerning Aviation Education at Kent State University," Dec. 12, 1989, private collection.

36. Beckwith to Moore, letter, Dec. 12, 1989, Schwartz papers, KSUSCA; Beckwith, handwritten notes, Jan. 2, 1990, private collection; Rebecca Buckingham, "Plan to Preserve Flight Program Offered to KSU," *DKS,* Jan. 16, 1990.

37. OCAE, organizational document, n.d., private collection; OCAE Roster of Officers, Committee Chairs, n.d., private collection.

38. Wilbur Biggin, letter to the editor, *DKS,* Nov. 30, 1989; Maya Eckstein, "Money Considered in Flight Decision," *DKS,* Nov. 29, 1989; Schwartz and Moore to Wilbur Biggin, letter, Dec. 1, 1989, Schwartz papers, KSUSCA.

39. School of Technology Faculty to Moore, Dec. 7, 1989, Schwartz papers, KSUSCA; Moore to School of Technology Faculty, Dec. 13, 1989, Schwartz papers, KSUSCA.

40. Schwartz to John Glenn, letter, Dec. 15, 1989, Schwartz papers, KSUSCA; Schwartz to W. H. Traub, letter, Dec. 19, 1989, Schwartz papers, KSUSCA.

41. Mary O'Brien, "Students Enlisting Aid of Alumni in Attempt to Save Flight Program," *DKS,* Mar. 7, 1990. The ads appeared in the *DKS* on Feb. 28, Mar. 2, Mar. 7, Mar. 8, and Mar. 15, 1990.

42. OCAE, open meeting announcement, *DKS,* Mar. 8, 1990.

43. David Dennis to Beckwith, letter, Jan. 31, 1990, private collection; Phil Trexler, "Grant Debts Could Reinstate Flight Program," *DKS,* Apr. 20, 1990; Phil Trexler, "Trustees Start Search for New KSU President," *DKS,* Apr. 24, 1990.

44. Rebecca Buckingham, "Schwartz Reopens Debated Program," *DKS,* May 2, 1990; Rebecca Buckingham, "Reasons Outlined in Report of Actions," *DKS,* May 2, 1990; Phil Trexler, "Viewpoint," *DKS,* Sept. 11, 1990.

45. Phil Trexler, "KSU's Ninth Leader to Step Down in Dec. After Eight-Year Reign," *DKS,* Mar. 21, 1990; Maya Eckstein, "Budget, Admissions for Flight Program Down," *DKS,* Aug. 29, 1990.

46. Craig Muder, "Training Planes Involved in Mid-Air Collision," *DKS,* Oct. 23, 1990; Terri Palmer, "Collision Forces Procedure Changes at Airport," *DKS,* Oct. 24, 1990.

47. "Profile: Sarah Deal, U.S. Marine Corps," *ABJ,* Mar. 29, 1999; "In the Region," *ABJ,* Dec. 4, 1999; "Lt. Col. Sarah Deal, '92, First Female Marine Aviator," Kent State University Alumni Association, accessed May 1, 2017, http://kentstate.imodules.com/s/401/social.aspx?sid=401&gid=1&pgid=2734; Shannon Huffman Polson, "She's Got Grit: The First Female Aviator in the USMC," accessed , June 6, 2016, http://aborderlife.com/grit-project-blog/shes-got-grit-the-first-female-aviator-in-the-usmc.

48. Tony Kennedy, "Northwest is Closing Gap in

Hiring of Women, Minorities," *Star Tribune* (Minneapolis, MN), July 6, 1998; Rachel Solomon, "Delta's First Black Female Captain: Taking off Was the Thrill of My Life," *Delta News Hub*, Feb. 9, 2017, https://news.delta.com/delta-s-first-black-female-captain-taking-was-thrill-my-life; "Captain Stephanie Johnson," *Organization of Black Aerospace Professionals*, accessed Apr. 13, 2017, http://www.obap.org/stephanie-johnson.

12. A New President, New Challenges, 1991–2006

1. Carol A. Cartwright, President's Message to the University Community, May 3, 1991, Cartwright papers, KSUSCA.

2. Eugene G. Ripple to Cartwright, memorandum, July 26, 1991, private collection; Eugene G. Ripple, "President's Report for the Kent State University Airport," July 25, 1991, private collection.

3. Eugene G. Ripple, "President's Report for the Kent State University Airport," July 25, 1991, private collection. The reporting line of the airport to the Business Affairs Division remained. The only changes were in 1995–96 when the airport became part of the transportation services unit and then in 2000 when it became part of a new public safety unit within the Business and Finance Division.

4. Jean Angelo, "Director Retires but Doesn't Plan to Quit Working," *DKS*, Mar. 4, 1992; Jennifer R. Barrett, "Airport Renovations Expected to Increase Revenue," *DKS*, Oct. 15, 1992.

5. Ripple, "President's Report for the Kent State University Airport," July 25, 1991, private collection; John Duncan (KSU aeronautics faculty member), phone conversation with the author, May 3, 2018.

6. Jeffrey J. Guritza, "Kent State Airport Director's Memories Take Flight Again," *DKS*, July 14, 1993.

7. Jackie Meeuwen, "Police, FAA Seek Plane Stolen from KSU Airport," *ABJ*, Aug. 12, 1993; Angela Gent, "Plane Disappears from Airport," *DKS*, Aug. 18, 1993; Kent State University Police Department, Incident Report 9304310, Aug. 10, 1993.

8. Kelley Foley and Beth M. Gorczyca, "Students Fear Flight Program May Be Cut," *DKS*, Apr. 27, 1994.

9. Enrollment had gone from 405 flight students in 1987 to 202 in 1993. Flight hours decreased from 19,001 in 1987 to 7,520 in 1993. "School of Technology: Center for Aviation Studies," Mar. 21, 1994, private collection.

10. Jennifer L. Boyett, "Airport Director Loses His Job without Notice," *DKS*, June 29, 1994; Jennifer L. Boyett, "Flight Program Stays, KSU Officials Announce," *DKS*, June 29, 1994; Cartwright to KSU Airport and Aerospace Technology Task Force, memorandum, July 8, 1994, Cartwright papers, KSUSCA.

11. KSU Airport and Aerospace Technology Task Force, Report, Feb. 23, 1995, private collection.

12. KSU Airport and Aerospace Technology Task Force, Report, Feb. 23, 1995, private collection.

13. University News and Information, "Kent Studies Flight Program, Opens New Regional Technology Links," Jan. 1996, private collection; Jennifer Moseley, "Flight Minor Available in the Fall," *DKS*, Apr. 16, 1996; Julie Vennitti, "Flight Classes Opened to All Majors," *DKS*, July 24, 1996; Jennifer Smith, "Pilot License Available through Flight Program," *DKS*, Sept. 5, 1996; Aaron D. Morrow, "Flight Made Easy with Continuing Studies Program," *DKS*, Oct. 15, 1999. A proposal also came from Commercial Aviation Corporation, the airport's fixed-base operator (FBO). Its president, Albert Beckwith, wished to pursue a land lease agreement to construct T-hangars, providing revenue for his business as well as rental income for the university. Although Beckwith's plan generated no interest, he would persist and put forward other proposals in the future.

14. KSU Trustees, meeting minutes, Feb. 2, 1995, BOT, KSUSCA; Kent State University Airport, Annual Report, 1995–1996, private collection; Beckwith to Diane Johnson, letter, Mar. 21, 1995, private collection.

15. Nicole Wisniewski and Vera Fedchenko, "Cartwright Receives Raise," *DKS*, Nov. 30, 1995.

16. "The 1990 Count: Final Census Results," *ABJ*, Jan. 25, 1991; "Population up in Tri-County," *ABJ*, Nov. 6, 1971. Stow was unincorporated until 1956, when it became a village; it became a city in 1960. "12,227 in Stow, Gains City Status," *ABJ*, May 14, 1960; Maura McEnaney, "Trotting out a Housing Development," *ABJ*, Aug. 19, 1993; Mary Ethridge, "New Shopping Plaza May Help Old One," *ABJ*, Mar. 9, 1994. New housing developments included River Park Estates (1978, Munroe Falls), Partridge Highlands (1979, Stow), and Hunter's Crossing (1987, Stow).

17. Kent State University and NBBJ, "Kent Campus Master Plan, the Kent Meadows," June 1996, 51–54, KSUSCA.

18. Kent State University Airport, Annual Report, 1995–1996, private collection; Kent State University Airport, Annual Report, 1996–1997, private collection; Tom Grossman (former KSU flight-training director), phone conversation with the author, Oct. 30, 2017.

19. Rick Neale, "KSU Airport Seeks FAA Grant to Help Fund Runway Renovations," *DKS*, June 24,

1998; KSU Trustees, Resolution, Mar. 10, 1999, BOT, KSUSCA.

20. Daniel Brindle, "Five New Planes Added to Kent Fleet," *DKS*, Feb. 10, 2000; Kent State University, "Statement Regarding the Kent State Aeronautics Program," *DKS*, Mar. 8, 2001; "Continental Express Launches Ab Initio Training at Kent State," *Aviation Week Network*, Apr. 2, 2001, http://aviationweek.com/awin/continental-express-launches-ab-initio-training-kent-state.

21. Erin Antognoli, "Soaring Past the Stereotypes," DKS, Jan. 18, 1996; Donn Gaynor, "Flight for Freedom Is Theme of Kent Women in Air Race," *ABJ*, June 22, 1969; Luella Cordier, "Angel Derby Next for Kent's Famed Aviatrix, Ruth Sitler," *R-C*, Apr. 24, 1970; Katie Johnson, "Retiring Professor Pioneer for Female Pilots," *DKS*, July 31, 2002.

22. Valli Jo Mullen, "Flight Ban, Attack Hurt KSU Flyers," *DKS*, Sept. 13, 2001.

23. Richland Engineering Ltd., Kent State University Master Plan Report, draft, June 2004, revised Nov. 2006, https://ksuairportplan.files.wordpress.com/2013/03/2006-master-plan-report-no-appendices1.pdf, Executive Summary.

24. "Kent State May Close Stow Airport," *ABJ*, Aug. 20, 2002.

25. Beckwith to Tom Clapper, email communication, Oct. 9, 2002, private collection; "IMPORTANT! Please Sign This Petition," 2002, private collection; "Outline re the Airport in Stow," Sept. 14, 2002, private collection; Deborah Guziak, "Flight School Owner Questions KSU's Plans," *R-C*, Sept. 19, 2002; Rudy Rudolph to Beckwith, email communication, Sept. 9, 2002, private collection; Beckwith to Stow Council Office, email communication, Oct. 9, 2002, private collection.

26. Jessica Alaimo, "Kent State Airport Faces Questionable Future," *DKS*, July 23, 2003.

27. "Notice of a Public Meeting," Dec. 10, 2003, private collection; Richland Engineering Ltd., Kent State University Master Plan Report.

28. Tracy Wheeler, "KSU Gets Earful on Future of Its Airport," *ABJ*, Jan. 22, 2004; Leah Campanalie, "Airport Reports Future Plans," *DKS*, Feb. 20, 2004.

29. Elbert Starks III, "Friends of KSU's Airport Make Case for Keeping It," *ABJ*, Mar. 24, 2004; Stow-Munroe Falls Chamber of Commerce, Board of Directors, Resolution, Apr. 15, 2004, private collection; Joe Durbin, "Petitions Oppose Airport Expansion," *ABJ*, Apr. 7, 2004; Stephen Dyer, "Neighbors Fear KSU Runway Plan," *ABJ*, Mar. 6, 2004.

30. Paul L. Gaston to David Creamer and Shirley Barton, memorandum, Feb. 2, 2004, private collection; Aeronautics Division to Shirley Barton, Response to Provost Gaston Questions, Mar. 10, 2004, private collection.

31. Lindsay Wargo, "University to Close Kent State Airport," *DKS*, Apr. 13, 2004; Leah Campanalie, "KSU Decision to Close Airport in Limbo," *DKS*, Apr. 14, 2004; Irene R. Porter to Cartwright, May 21, 2004, private collection; Andrew Schunk, "FAA Tells KSU: Airport Closure 'Highly Unlikely,'" *Stow (OH) Sentry*, June 6, 2004.

32. KSU Trustees, meeting minutes, May 25, 2004, BOT, KSUSCA; Andrew Schunk, "Airport in FAA Hands," *Stow (OH) Sentry*, May 30, 2004.

33. Stephen Dyer, "Kent State Stands Pat on Closing of Airport," *ABJ*, July 15, 2004; Andrew Schunk, "Kent State to Stay the Course with Airport Proposal," *Stow (OH) Sentry*, July 18, 2004; "Stow Council Discussion on Airport Gets Heated," *ABJ*, July 23, 2004.

34. "Save This Airport," petition, Sept. 12, 2004, private collection.

35. Julie Vennitti, "Aviation Day for the Public," *DKS*, Aug. 7, 1996.

36. Aaron D. Morrow, "History Takes Flight at Aviation Day '99," *DKS*, Sept. 14, 1999; Madelin Esquivel, "Aviation Day Flies High with Enthusiasm," *DKS*, Sept. 8, 2003.

37. "Flapjacks, Flight and Festival Fun!," advertisement, *DKS*, Aug. 22, 2001; Joel Tianello, "Taste of Stow, Aviation Day Prove Fun," *DKS*, Sept. 11, 2001; Dennis Baden, email message to the author, Feb. 24, 2017.

38. Dennis Baden, email message to the author, Feb. 24, 2017.

39. Kristin Casale, "Volunteers Needed for Aviation Heritage Fair," *Stow (OH) Sentry*, Aug. 6, 2006.

40. David B. Burroughs, letter to the editor, *DKS*, Feb. 3, 1995; Joshua Hudson, "Taking Off," *DKS*, Oct. 8, 2004; David Carr, "After 19 Years, Flight Team Is Set to Compete Nationally Once Again," *DKS*, Oct. 21, 2005.

41. Holly Mueller, "Aeronautics Becomes Only Accredited Program in Ohio," *DKS*, Apr. 10, 2006.

42. Tyrel Linkhorn, "State-of-the-Art Planes for KSU," *DKS*, Jan. 19, 2006; Mueller, "Aeronautics Becomes Only Accredited Program in Ohio"; Rebekah Mosora, "KSU's School of Technology to be a College," *DKS*, Dec. 9, 2005; "Kent State University," Aviation Accreditation Board Inter-national, accessed Sept. 5, 2016, http://www.aabi.aero.

43. Leslie Arntz, "Cartwrights in Flight," *DKS*, Nov. 29, 2006.

13. Reaching the Century Mark

1. David Poluga (KSU Airport manager), email message to the author, June 26, 2017.

2. Albert Beckwith notes on meeting with dean Isaac Nettey, Jan. 29, 2007, private collection; Beckwith to AOPA, "1G3 Kent State University Airport ALERT," email correspondence, Jan. 29, 2007, private collection; Rick Annon, "Airport Expansion Plans Hit Some Turbulence in Portage," *ABJ*, Mar. 14, 2007; "Portage County Airport and Your Dollars," *PCCAG Newsletter*, May 8, 2007, private collection.

3. Beckwith to Irene Porter, email communication, Oct. 19, 2007, private collection; Dino DiSanto to Beckwith, email communication, Feb. 6, 2008, private collection.

4. Poluga, email message to the author, Sept. 18, 2018; Katherine S. Jones to Thomas J. Euclide, letter, Apr. 15, 2008, private collection.

5. Poluga, email message to the author, June 26, 2017.

6. C&S Engineers, Inc., *Kent State University Master Plan, Final Report* (Cleveland, OH: C&S Engineers, 2016), ES:1.

7. Alice Dunaway, "Kent State Airport Receives $600,000 Grant for Analysis," *DKS*, Aug. 29, 2012; C&S Engineers, *Kent State University Master Plan*, ES:2–3; Commercial Aviation Corp., "Suggested 2013 Proposed Master Plan Improvements: Permanent Structures," 2013, private collection; Andrew Bugel, "Kent Airport Remembers Long-Time Tenant and Mentor," *KentWired.com*, Oct. 20, 2014, http://www.kentwired.com/latest_updates/article_5267c984-58aa-11e4-b91e-001a4bcf6878.html.

8. Audit and Compliance Committee, Board of Trustees meeting, Kent State University, Mar. 13, 2013, BOT, KSUSCA; "FAA Provides Grant Fund to Kent State Airport for Sustainability Efforts," *R-C*, Sept. 20, 2013; "Airport Sustainability," Federal Aviation Administration, https://www.faa.gov/airports/environmental/sustainability, May 11, 2017; Poluga, email message to the author, June 23, 2017.

9. Alicia Balog, "Lefton Announces Retirement," *DKS*, Apr. 18, 2013; Andrea Delph, "New Aeronautics Building Leaves Room for Growth," *DKS*, Jan. 22, 2015.

10. C&S Engineers, *Kent State University Master Plan*, ES:14; C&S Engineers, *Kent State University Master Plan*, ES:6–7. President Warren announced her retirement, effective July 1, 2019. Henry Palatella, "KSU President Warren Announces She Will Step Down," *KentWired.com*, Oct. 23, 2018, http://www.kentwired.com/latest_updates/article_ade6a286-d6e0-11e8-8bc6-3f208e65be5f.html.

11. C&S Engineers, *Kent State University Master Plan*, ES:8; Andrew Bugel, "Aeronautics Program Seeks Options to Improve Runway, Airport Hangar," *DKS*, Nov. 24, 2014; C&S Engineers, *Kent State University Master Plan*, ES:9.

12. C&S Engineers, *Kent State University Master Plan*, 2:63; "Kent State University Airport Reaches Out to the Community," *eInside*, Apr. 25, 2017, http://einside.kent.edu; "Kent State University Airport Hosts Props-4-Paws Pet Supply Donation Event," *eInside*, Sept. 16, 2016, http://einside.kent.edu.

13. C&S Engineers, *Kent State University Master Plan*, ES:15; KSU Trustees, Resolution, Sept. 9, 2015, BOT, KSUSCA.

14. Poluga, email message to the author, June 27, 2017.

15. C&S Engineers, *Kent State University Master Plan*, 6:8–10. The AWOS was installed and activated on April 4, 2019. The 1G3 airport identifier was maintained as part of the airport's history. Poluga, email message to the author, May 6, 2019.

16. C&S Engineers, *Kent State University Master Plan*, 6:11.

17. Natalie Meek and Colin Baker, "$6.5 Million Building at University Airport Promises More Space," *DKS*, Sept. 25, 2017; Poluga, discussion with the author, June 23, 2017.

18. "Kent State Renames Buildings to Recognize FedEx, Outgoing President," *ABJ*, May 9, 2019; "KSU Receives $1.5 Million Gift from FedEx," *R-C*, Sept. 29, 2016; "Kent State and FedEx Announce $1.5 Million Donation that Propels Aeronautics to Greater Heights," *eInside*, Sept. 30, 2016, http://einside.kent.edu; "Board Approves New Airport Classroom Building to Support Growing Aviation Program," Kent State University News, Mar. 7, 2018, https://www.kent.edu/kent/news/success/kent-state-university-board-trustees-approves-10-year-plan-reshape-revitalize-kent-campus.

19. C&S Engineers, *Kent State University Master Plan*, 6:16–18.

20. "KSU Aeronautics Fair Offers a Day of Flight, Education and Fun," *Stow (OH) Sentry*, Sept. 4, 2016; Lindsay Ridinger, "Fair for Flight," *KentWired.com*, Sept. 11, 2009, http://www.kentwired.com/latest_updates/article_b5b25bf6-ae4d-5367-adb6-1d1b396c0c61.html.

21. Faith Cook (Stow-Munroe Falls Kiwanis Club), email message to the author, Sept. 18, 2017; "Kent State University Aeronautics Fair," College of

Aeronautics and Engineering, accessed Sept. 1, 2017, http://www.kent.edu/AeroFair.

22. Faith Cook (Stow-Munroe Falls Kiwanis), email message to the author, Sept. 18, 2017; "Kent State University Aeronautics Fair."

23. Abbey Linville, "Precision Flight Team Wins Conference," *KentWired.com,* Nov. 14, 2008, http://www.kentwired.com/latest_updates/rticle_65afd4ef-b14e-5234–8635–02f685cfcd36.html; Leighann McGivern, "KSU Precision Flight Team Secures National Championship," *DKS,* Apr. 18, 2011; "SAFECON Loening Trophy Judging Criteria," National Intercollegiate Flying Association, accessed May 29, 2017, https://www.nifa.us/loening.pdf; Matt Fredmonsky, "Kent State Flight Team Recognized at Statehouse," *KentPatch* (Kent, OH), Nov. 27, 2011, https://patch.com/ohio/kent/kent-state-flight-team-recognized-at-statehouse.

24. Abbey Linville, "A Donation on the Fly," *KentWired.com,* Nov. 14, 2008, http://www.kentwired.com/latest_updates/article_0998e0e6–514c-508a-9d5c-dd95f48d0411.html.

25. Seth Cohen, "'Prince of a Pilot' Honored at Memorial," *DKS,* Apr. 4, 2011.

26. Kent State Accreditation, Aviation Accreditation Board International, accessed Sept. 17, 2017, http://www.aabi.aero/programs.html; "Aeronautics Programs Enrollment and Graduation Data," KSU College of Aeronautics and Engineering, accessed Sept. 17, 2017, https://www.kent.edu/cae/aeronautics-programs-enrollment-and-graduation-data.

27. Katie Byard, "KSU Tuition Won't Increase, *ABJ,* June 1, 2017; Robert G. Sines Jr. (interim dean, College of Aeronautics and Engineering Education), email message to the author, Sept. 7, 2017.

28. "Kent State University Airport Named Ohio Airport of the Year," *ABJ,* June 1, 2017; "OAA Annual Aviation Awards," accessed June 1, 2017, http://www.ohioaviation.org.

29. Poluga, discussion with the author, Oct. 2, 2017.

Epilogue

1. Today, Kent State University, Bowling Green State University, Ohio State University, and Ohio University have active collegiate aviation programs. Miami University does not.

Index

Page numbers in italics refer to illustrations.